Empirical Methods in Language Studies

# ŁÓDŹ
# STUDIES IN LANGUAGE

Edited by
Barbara Lewandowska-Tomaszczyk
and Łukasz Bogucki



**Vol. 37**

PETER LANG
EDITION

Krzysztof Kosecki / Janusz Badio (eds.)

# Empirical Methods in Language Studies

PETER LANG
EDITION

**Bibliographic Information published by the Deutsche Nationalbibliothek**
The Deutsche Nationalbibliothek lists this publication in the Deutsche Nationalbibliografie; detailed bibliographic data is available in the internet at http://dnb.d-nb.de.

**Library of Congress Cataloging-in-Publication Data**
Empirical methods in language studies / Krzysztof Kosecki ; Janusz Badio (eds.)
pages cm. — (Lodz Studies in language; Vol. 37)
ISBN 978-3-631-65664-8 (Print) — ISBN 978-3-653-04976-3 (E-Book) 1. Applied linguistics—Research—Methodology. 2. Language and languages—Research—Methodology. I. Kosecki, Krzysztof, editor. II. Badio, Janusz, editor.
P129.E47 2015
407.2'1—dc23

2015022224

This publication was financially supported by
the Institute of English Studies of the University of Łódź.

ISSN 1437-5281
ISBN 978-3-631-65664-8 (Print)
E-ISBN 978-3-653-04976-3 (E-Book)
DOI 10.3726/978-3-653-04976-3


Peter Lang Edition is an Imprint of Peter Lang GmbH.

Peter Lang – Frankfurt am Main · Bern · Bruxelles · New York · Oxford · Warszawa · Wien



This publication has been peer reviewed.

www.peterlang.com

# Table of contents

# Preface

The present volume contains 22 papers divided into four sections. The first section concentrates on experimental and survey methods, and the five papers explore the problems of events and their coding into sentence format in English and Polish, genericity, linguistic worldview, signed metaphor comprehension, and second language acquisition. Part two, the largest in the volume, contains nine papers that rely on language corpora to shed light on various aspects of grammar, conceptualization, emotions, time, gestures, and discourse. The papers in part three focus on analysis of linguistic expressions related to literature, semantics, and philosophy. Part four contains five papers which either employ or exemplify miscellaneous methods, for example experiment, language corpora, and converging evidence from linguistic examples, in the description of iconicity, literary texts, second language learning, translation, and meaning construal.

# Part One: Experimental and Survey Methods

Janusz Badio
*University of Łódź, Poland*

# Events and sentences in story construal of English-native and Polish-foreign language users: experimental methodology and outcomes

**Abstract:** The article presents an experimental study and its three constituent analyses of narrative events and their construal-coding in sentence format of a story. The theoretical position that provides the starting point for the research questions posed in this work is traced down to Chafe (1994), who concluded that, especially in writing, the sentence can be understood as a centre of interest with scope often much larger than a single focus of consciousness. The author investigates how Polish-native and English-foreign language users construe and code events presented in a children's cartoon story. The article presents the materials, questions, descriptive statistics, hypotheses, and inferential tests together with some follow-up qualitative observations of the written interpretations of the story. In general, the findings support the more general hypothesis (cf. Badio 2014) that the subordinate goal of construal and coding operations is to properly adjust the salience of entities (e.g., objects, characters, locations, reified processes) and relations (e.g., features of objects, processes). It is, however, the methods used in this study that are highlighted, in agreement with the general theme of the present volume, which is empirical methodology in studying language.

Keywords: story, event, construal, coding, ANOVA, Chi-square, t-test, sentence.

## 1. Introduction

This work describes a part of a larger project (Badio 2014) on construal and coding of narrative events. Its foremost hypothesis stated that achieving cognitive prominence of selected entities is the language users' superordinate goal. The empirical studies in this work looked at ways events are perceived and delineated from a stream of video, whether their linguistic coding is isomorphic to their perception and memory, or how they are selected to construe the causal structure of a story neutrally or in such a way as to blame one participant for what has happened. Events' natural habitat is a story, or a narrative, that is, larger discourse coherence. Construal operations refer both to the level of a single event and discourse. It is the discourse level, the whole story level, which constitutes the research focus presented in this article. Three related general research questions were asked:

a) How will the participants construe and code a cartoon story presented in the form of a silent video, as opposed to a static sequence of pictures (screenshots) from the same video?
b) What sentence types are preferably used by native-Polish and English-foreign language users?
c) How will the participants in the study package events and their parts in sentence format?

As already mentioned, the participants ($N = 30$) retold a cartoon story in writing. Data collected in this way were subsequently coded in the format of a Microsoft Access dBase with three tables: SENTENCES, PARTICIPANTS, and CONDITIONS. The first of these tables contained such descriptors of the sentence unit as: the actual sentence, type (clause, coordinate, subordinate, complex), number of events coded in a sentence, and other comments. The table holding information about the participants contained only the number that uniquely identified them, whereas the conditions table included all the conditions (independent variables), in which the story descriptions took place. Importantly, the main sentence table was related to the participants table with the relation one-to-many, i.e., one participant was referred to all the sentences s/he produced in writing, and the same was true of the relation between the condition and sentence tables, i.e., one experimental condition stored all and only those sentences that were produced in it. The format of a relational dBase allowed multiple ways of accessing the created corpus of texts.

Before the presentation of the actual study with its analyses, some theoretical background is offered in section two.

## 2. Theoretical Background

Construal constitutes the heart of Cognitive Linguistics' theorizing about language. Its theoretical foundations were laid by Langacker (e.g. 1987, 2008), Talmy (2000 and earlier), or Croft and Cruse (2004). Perhaps its most influential definition was offered by Langacker (ibid.), who claimed that construal is a relation between the conceptualizer and the object that is conceptualized. This relation is described in the works of the above linguists under different, more detailed categories of: selection, perspective, abstraction, prominence, specificity (Langacker ibid.), schematization, perspective, attention, force dynamics (Talmy ibid.), and attention, judgment/comparison, perspective, constitution/gestalt, force dynamics (Croft and Cruse ibid.). Despite the differences in categorizing construal operations, their underlying common denominator is the belief that it is the so-called cognitive salience/prominence, under the control of attention and its multiple strands[1] that is the super-ordinate goal of any human attempt to construe some content and linguistically code it.

An event, one of the key units, is defined here after Zacks and Tversky (2001: 7) to be a "segment of time at a given location perceived by an observer to have a beginning and end". Its perceptions are correlated with specific linguistic coding in different languages. Moreover, events, by analogy to physical objects, contain parts, and they also come in variable types. They frequently involve some evolution through time of a certain state of an object, or a set of objects in the case of complex scenes, and they are typically, though not necessarily, coded by verbs in English (e.g. *He jumped* v.s. *He gave a jump*). Any fragment of the conceptual content of an event (time, space, manner, etc.) can, in theory, be focused on and coded with an appropriate linguistic unit.

The sentence is also important for this work because after Chafe (1994), it is understood to code the so-called *centre of interest*, a human attempt to expand the limited scope of the active focus of conscious experience. The following discourse exchange (diacritics provided JB)[2] illustrates the idea.

1) a) .. I was on the ^bus ^toda=y,
b) .. ^a=nd there was this w^oman s^aying,
c) .. that her s^on,
d) .. ^works .. for the r^anger s^ervice or whatever.
e) .. and .. there was sn^o=w,
f) .. ch^est high,
g) .. at Tu^olome M^eadows. Chafe (1994)

In speech, this fragment consists of [a-g] intonation units, each unit coding the content of a separate, changing focus of attention. However, units [a-d], and [e-g] form larger coherences that could be coded by only one sentence in writing each, *I was on a bus today, and there was this woman saying that her son works for the ranger service*, and *There was snow chest high at Tuolome Meadows,* accordingly.

1 Readers who are interested in different models and theories of *attention* are referred to Nęcka et al. (2006). Here this term is understood after (Baars 1997; Chafe 1994, 1996, 1998, 2003) as a set of implicit processes that give rise to conscious experience.

2 [..] – pause, [^] – sentence stress; [=] – anacrusis (vowel lengthening); [,] – end of intonation unit with pitch indicating that the idea is not finished; [.] – end of intonation unit.

Moreover, it is important to explain that the story retellings used in this study come from a native form of language (Polish), as well as a foreign language (English). With interest on selected aspects of processing, rather than formal aspects of grammar and vocabulary, this was possible following the so called psychological tradition in SLA (cf. Hulstijn (2007: 197). While the linguistic tradition places emphasis on linguistic representations of knowledge, the psychological tradition, which this paper follows, accentuates the question of how learners process information in a foreign language. In agreement with the psychological tradition, the present work argues that a foreign language is an autonomous and legitimate object of research, though comparisons to a native language are expected and natural. Such a perspective is at least partly concomitant with research on second language variability, some aspects of transfer, cognitive descriptions of second language production, status of knowledge, and cognitive control over access to the first/second language, which also Ellis (2008) discusses.

Having introduced the key concepts of this study, i.e.: construal, coding, event and sentence, the author will proceed to the next part, where the materials, procedures and specific hypotheses together with the remaining, and more detailed theoretical points, will be explained.

## 3. Materials and procedures

The video selected for this study is the *Bolek and Lolek: First day of summer holidays*[3], a popular children's silent cartoon in Poland. The choice of a cartoon story was dictated by its accessibility on internet, but also the observation that it shows clearly the characters and their actions. Only the main characters are shown without unnecessary clatter, which is another advantage, and they are contrasted well against the background. The actions performed by the boys are easy to interpret, and the plot is simple to follow.

In one group the viewers were asked to watch the video with sound off in the classroom on TV, while the participants of the experiment in the other group were only shown the screenshots from the video arranged chronologically. Both groups first spent the same amount of time viewing their respective inputs, and then they had to retell the story in writing. All participants were granted fifteen minutes to finish this writing task. The instructions provided on their worksheets in Polish were translated by the author into English, and they are presented below:

3 The cartoon was directed by Władysław Nehredecki and illustrated by Alfred Ledwig (1974–1980).

Stage 1 – watching
You will watch a silent video, 8 minutes and 58 seconds long. Watch carefully. After watching only once, please write in Polish [the other group, "write in English"] what happened in the story. The story must be written in the 3rd person. How long the written retelling should be, and what to include depends on your choice. The point is to recreate in writing the events presented in the film.

Stage 2 – writing
Write the story. You have 15 minutes to complete the task, and there is no word limit.

The instructions for those who watched the screenshots from the same film were as follows:

Stage 1 – watching
You will watch screenshots from a silent video, 8 minutes and 58 seconds long. You will have 8 minutes and 58 seconds to look at the pictures arranged chronologically from left to right and top to bottom. After watching the pictures, please write in Polish [the other group, "write in English"] what happened in the story. The story must be written in the 3rd person. How long the written retelling should be, and what to include depends on your choice. The point is to recreate in writing the events presented in the film.

Stage 2 – writing
Write the story. You have 15 minutes to complete the task, and there is no word limit.

The participants of the study were all advanced / proficient students of English as a foreign language, all during their first or second year at M.A. level of English Philology in Łódź, Poland, and they agreed voluntarily to take part in the experiment for partial credit on their seminar course.

## 4. Analysis one – sentences in story retellings

This analysis, i.e., the first part of the cartoon story retellings experiment, had the goal of answering the following research question. Will speakers of Polish-native and English-foreign languages code events presented by either a video or picture sequence differently? Will the four independent groups (i.e., PL-Vid, PL-Pic, Eng-Vid, Eng-Pic) differ as regards the number of sentences they will decide to code in writing. As one can see, apart from language, the other independent variable was input type (continuous/dynamic-video or static-pictures). It was predicted (cf. Skehan 1998) that different task types could affect language production. Here, the choice of continuous video rather than the static sequence of screenshots from the same video was predicted to result in a larger number of centres of interest, and thus also sentences because, in theory, there is no limit to what can attract the viewers' attention during a continuous stream of video. Also, it was predicted that the Polish-native language users would produce more sentences (centres of interest), as one can expect, the use of one's native language should pose fewer language coding problems.

A little repetitively perhaps, the following hypotheses were formulated as regards the two experimental conditions:

a) Video-Pictures

H1: The participants in the video condition will produce more sentences than in the picture condition.
H0: There will not be significant differences in the numbers of sentences between the languages in this respect.

b) Polish-English languages

H1: Polish stories will contain more sentences than their English counterparts.
H0: There will not be a statistically significant difference between the numbers of sentences produced in the written descriptions of the input story as a function of language choice.

c) Interaction
Moreover, it was predicted that there may be interaction between the choice of language and input type, such that the choice of the native language coupled with the video input should lead to a greater number of sentences in the written interpretations of the input story.

Hence, the whole design was an ANOVA 2x2 because two independent variables were tested, language and input type, each variable with two levels represented by the choice of either Polish or English, and Video or Static Pictures (screenshots from the same video).

The analysis of descriptive statistics showed that the largest mean number of sentences was produced in the PL-Vid condition, followed by Eng-Pic, Eng-Vid and last, PL-Pic conditions. It was also apparent that the Polish language condition rendered more sentences per retelling ($M$-PL = 16.75) compared to the choice of English ($M$-Eng = 13.71). The comparison between the video and picture conditions showed that the former (video) condition invoked more (M-Vid = 17.53) sentences than the picture condition ($M$-Pic = 13.30).

So that these results could be generalised, ANOVA 2x2 was run with the following table presenting the important parameters:

*Table 1: ANOVA results of the study of sentences in pictures/video vis a vis Polish/English conditions*

| ANOVA Summary | | | | | |
|---|---|---|---|---|---|
| **Source** | **SS** | **df** | **MS** | **F** | **p** |
| Rows – language: Polish / English | 68.81 | 1 | 68.81 | 2.47 | 0.1281 |
| Columns – input: Picture / Video | 145.2 | 1 | 145.2 | 5.2 | 0.031 |
| Language x input interaction | 307.01 | 1 | 307.01 | 11 | 0.0027 |
| Within groups variation: (i.e. error due to individual variation) | 725.65 | 26 | 27.91 | | |
| Total | 1246.67 | 29 | | | |

A significant main effect was obtained only for input type ($F$ (1, 26) = 5.2; $p < .05$). ANOVA further revealed a significant interaction between language and input type: ($F$ (1, 26) = 11; p<.05). However, there was no significant main effect for language choice: ($F$ (1, 26) = 2.47; $p > .05$).

To sum up, language choice was not connected with a difference in the number of sentences the writers used to retell the story, but the selection of video over pictures led to their significantly greater numbers. There was also a robust interaction between language choice and video; in was in the video condition, where the writers used Polish (their native language), that resulted in a greater number of sentences.

## 4.1 Discussion of results

Language choice had no effect on the number of written sentences. The participants reported the input story to be relatively easy to write about, and also that they did not avoid coding in language any specific aspect of the input story. In their post-task commentaries collected immediately after the main task, they also stated that there were problems connected with coding the details of the scenes around the episode of car repair. This shows that time is not a good memory index. The details did not form a cause-effect chain, which normally leads to enhanced memory of events (cf. Barsalou 1998).

Moreover, though the instruction required the participants to retell the story, many of them stated the task required creativity. It was especially effortful in the "picture" condition, as the static pictures disturbed the normal continuity presented on the video, and the writers in this condition must have been compelled to impose interpretations on how to connect the pictures to form some coherent whole. The effect of language choice (Polish) surfaced only when it was coupled with the video presentation, and hence interaction was reported (see above). The next section presents the report of the second analysis into the type of sentences and events used in the study.

## 5. Analysis two – sentences in Polish vs. English story retellings

The rationale for analyzing the numbers of different types of sentences in Polish-native and English-foreign retellings of the story was the contention that they code an event or a set of events in different ways. Single clauses typically contain one finite form of the main verb together with arguments that align with the trajector (primary, most salient figure), and one or more landmarks depending on the situation and the choice of the verb. Seven types of the so-called basic syntactic patterns (Quirk et al. 1972: 172) represent variable types of events. These patterns (SV, SVO, SVC, SVA, SVOO, SVOC, SVOA) contain obligatory elements: subject, objects, verb, complements and modifiers. A coordinate clause contains two cognitively autonomous contributory clauses, and their independence is correlated with equal cognitive salience of their profiles (imposed by verbs). As is well known, there also exist subordinate sentences, which contain one dominant clausal profile, with at least one or more other dependent ones expressed by the so-called dependent clauses, as well as more complex structures. The list below was used to categorize the sentences in the corpus.

a) single, independent finite clause
b) compound/coordinate sentence – two or more coordinate clauses
c) coordinate and subordinate sentences – sentences related in both ways

One can say that the criteria used in the task of coding the sentences into the three categories are both structural and cognitive (conceptual, semantic).

In coding an event, it is the finite verb form that is most salient, with infinitives and participles occupying the middle position, followed by reified verbs, e.g. *a jump*. Let us consider a few examples from the corpus.

1) Walking to the bus stop, he wanted to get on the bus.
2) He had a plan → He would get on the bus → At that time he was walking.

Example [1] ranks the cognitive salience of contributory, and overlapping processes in such a way that it is the profile of *wanted* that is most salient, presumably followed by *to get* and last, *walking*. Certainly, a separate experiment would be needed to test this prediction. For lack of space, the reader is referred to Langacker (1991, 2008) or Badio (2014) for a more detailed analysis. In sum, syntax offers a myriad of conventional ways of adjusting the cognitive prominence of things and relations of a conceptualization. Some of them have been discussed above, though certainly there has been no time for a fuller treatment of this topic because the major focus of the present work in set on experimental design and method.

## 5.1 Questions, hypotheses and results in analysis two

The biggest cognitive load on the cognitive system is posed when a person plans to construe and code in language a complex scene/event. It is arguably smaller when s/he recruits a single clause with a finite verb for the task, and becomes more challenging with coordinate, and subordinate sentences. It was predicted that despite being proficient in the foreign language, the participants writing in English would favour single and coordinate structures, and that subordination as well as more complex structures would be preferred by those who were asked to write in Polish. In connection with this the following general question was asked, "Does the use of a foreign language induce such thinking for speaking (Slobin 1996, 2004) that leads to less complex conceptualizations, more easily codable in a foreign language?", and the narrower research question was, "Will the use of Polish-native and English-foreign languages be independent of the number of different sentence types in the data?" At the risk of becoming a little repetitive, the detailed hypotheses are presented again below in a more formal way:

H1: The Polish versions of the story will contain more simple clauses, and fewer subordinate and complex sentences, and there will be no difference between Polish-native and English-condition as regards the numbers of coordinate sentences (the last category is additive, relatively simple and should not be avoided by foreign language users).

H0: The null hypothesis claimed that the numbers of different types of sentences referred to above, even if different between experimental conditions, will not be attributable to the choice of language.

The coding of the sentence types was performed by two linguistics, who reached 95 % agreement. The following table presents the actual numbers of different sentence types in the two experimental conditions.

*Table 2: Raw numbers of sentence types in four experimental conditions*

| **language** | **sentence type** | | | |
|---|---|---|---|---|
| | **simple** | **co-ord** | **subord** | **complex** |
| Polish | 52 | 44 | 57 | 39 |
| English | 73 | 67 | 81 | 48 |

There turned out to be no significant relationship between language choice and the number of different sentence types in the data set of Polish and English video retellings in, $X^2$ (3, $N = 460$) = 0.553; $p > .05$. The null hypothesis was accepted in the Chi-squared test of independence. Last, the third analysis in this study was concerned with the numbers of events coded in the Polish and English samples.

## 6. Analysis three – events in sentences

The question asked in this part of the study was, "Are native speakers of Polish able to bundle more information about events in a sentence than non-native speakers who are students of English as a foreign language?" The related hypotheses were formulated as follows:

H1:
Polish-native language retellings of the cartoon presented by video and sequence of pictures will contain more events than in the case of English-foreign language condition per single retelling.

H0:
The null hypothesis predicts that the numbers of events across conditions will not differ in a statistically significant way per single retelling as a function of language (Polish vs. English).

The analysis required some method of counting events in sentences. The task was far from straightforward, and three coders worked together to reach 95 % agreement. An event was counted if a finite, non-finite, or reified form of verb was used. Moreover, sometimes an event was implied despite any direct coding with a verb form. The interested reader is referred to Badio (2014: 196–202) for more details. Here only a few sentences will be quoted to exemplify the process of coding.

3) The man [1] *agreed* and [2] *lent* them the car. → two events
4) The boys [1]*are* probably [1] *at school* because they [2] *have schoolbags on their shoulders.* → two events

5) They [1] *did not know* what [2] *to do* with their spare time and began [3] *to think* of possible [4] *trips* to go on. → four events
6) [1] *Nie zbudowali* w końcu łodzi, [2] *zobaczyli* mężczyznę z wędkami i [3] *zapragnęli* [4] *łowić ryby* → four events
   lit: 'They did not build the boat in the end, saw a man with fishing rods and wished they could go fishing.'
7) Bolek od razu [1, 2 – refers to the second clause too] ] *rozmarzył się* o pieniądzach, i co [3] *kupić*, zaś Lolek o [4] *naprawie auta* i [5] *przejażdżce* → five events
   lit: 'Bolek at once dreamt of money, and what to buy, whereas Lolek aborted repairing the car and going for a drive'.

The above sentences shed some light on how careful one must be during the coding process. The use of more than one coder helped achieve higher internal validity and reliability of the experimental design and test results.

## 6.1 Results of descriptive statistics and inferential tests

In the first step towards the discussion of the results, the descriptive statistics of central tendency and dispersion are provided. The mean number ($M$ = 38.2) of events coded in the Polish condition per sample turned out to be bigger than its equivalent for the English condition ($M$ = 27.3). The standard deviations for these two different sub-samples of the corpus exhibited rather large differences (*SD*-pl = 18.5; *SD*-eng = 7.8; *Var.*-pl = 343.4; *Var.*-eng = 61.5). Moreover, the skewness of the data for each sub-samples was: *Skewness*-pl = 0.8, and = 0.7 for the English sub-sample. The general rule of the thumb is to treat values smaller than minus one, or bigger than one as indicating that the data are not normally distributed. In the case of this analysis, it does not exceed these values, which means that a parametric test of inference can be used, in our case the *t*-test for independent samples.

The result was statistically significant ($t$ (23) = 1.89; $p < .05$) in a one-tail *t*-test, which indicates that the experimental hypothesis should be, though tentatively, accepted. The choice of Polish did cause that the speakers used more events in their retellings.

Though the *t*-test at the 5% level was only just significant for single clauses, it was unimportant for coordinate structures, though significant again for the subordinate and complex (subordinate and coordinate) sentences, which the author had predicted. It was the subordinate sentences, where the differences between Polish and English group of writers were the biggest in terms of the numbers of events per sentence type. The following table summarizes this presentation.

*Table 3. Events and sentences in English and Polish retellings*

| | simple | coord | sub | sub-cc | total |
|---|---|---|---|---|---|
| pl-events | 84 | 161 | 199 | 174 | 618 |
| pl-sentences | 73 | 67 | 81 | 48 | 269 |
| *Proportion e/s* | 1.15 | 2.40 | 2.45 | 3.62 | 2.29 |
| eng-events | 55 | 97 | 113 | 126 | 391 |
| eng-sentences | 52 | 44 | 57 | 39 | 192 |
| *Proportion e/s* | 1.05 | 2.20 | 1.98 | 3.23 | 2.03 |
| *t*-test sig. at *alpha* = .05 | yes/no (harldy) | no | yes | yes | yes |

The Polish-native writers were able to squeeze in more events into their sentences and stories, though especially in subordinate structures this effect was most robust.

## 7. Qualitative analysis of selected events – some observations

Some example analyses that are more qualitatively oriented are also reported here to illustrate the empirical methodology of looking at what may be termed multiple construals. First, the author decided to pick up the scene that the story begins with, i.e., BOLEK AND LOLEK WALK. The following two screenshots illustrate it.

*1. Bolek and Lolek WALK.*

*2. B SAYS GOODBYE to O.*

The sentences that were used to describe this sequence of events in the video condition were different than the ones in the pictures condition in that the latter were more interpretative. Consider the following data:

VID PL
spacerowali [were walking], odprowadzał [was seeing off], wracali [were returning], odprowadził [saw off]

VID EN
had time to relax, went out (from school), wanted to, came back, were excited

PICT EN
split up, graduated from school, started to have free time, walked around the street, decided to go, came back, were walking, left school, decided to meet

PICT PL
odprowadził [saw off], rozdzielili się [split up], spotkali (ponownie) [met (again)], umówili [arranged to meet], skończyli (zajęcia) [finished (classes)]

By contrast to the Vid-PL condition, the verbs in the Vid-Eng condition tend to be more general, and they represent a more schematic level of catergorization, e.g., *went out* and *came*. They also tend to designate a subjective evaluation rather than fact, e.g. *had time to relax, were excited*. Generally, though a little tentatively, one may speculate that the Polish stories reflected the use of conventional construal-codings, whereas their English counterparts probably turned towards the use of various communicative strategies to meet the task requirements. The Polish-native stories had more events, especially in the video condition, where they stand out as more detailed, and are delivered with the use of a greater variety of syntactic, sub-ordinate options, which corroborates the conclusions reached in the quantitative part (see above).

## 8. General conclusions

In sum, the results of the study reported in the present article provide support for the theoretical model proposed, namely, that proper adjustment of the cognitive prominence of entities (e.g., objects, characters, locations, reified processes) and relations (e.g., features of objects, processes) is the subordinate goal of the operations of construal. As for methodology, the author highlighted the stages of experimental design: general question, research question, formulation of experimental (H1) and Null (H0) hypotheses. As for data analysis, it was shown how to report descriptive statistics, and last but not least, how to go beyond the limited samples one studies and make generalizations with the use *ANOVA* 2x2, *Chi-square* test of independence and a *t*-test in order to study language at the level of discourse (here stories and their events) from the theoretical standpoints of Cognitive Linguistics.

## References

*Badio, J.* (2014). *Construal and Linguistic Coding of Narrative Events*. Łódź: Wydawnictwo Uniwersytetu Łódzkiego.

*Baars, B. J.* (1997). "Some essential differences between consciousness and attention, perception and working memory". *Consciousness and Cognition* 6, 363–371.

*Barsalou, L.* (1988). "The content and organization of autobiographical memories". In: U. Neisser and E. Winograd (eds.), *Real Events Remembered: Ecological and Traditional Approaches to the Study of Memory*. Cambridge: Cambridge University Press. 193–243.

*Chafe, W.* (1994). *Discourse, Consciousness and Time*. Chicago: The University of Chicago Press.

*Chafe, W.* (1996). "How consciousness shapes language". *Pragmatics and Cognition* 4 (1), 35–54.

*Chafe, W.* (1998). "Language and the flow of thought". In: M. Tomasello (ed.), *The New Psychology of Language: Cognitive and Functional Approaches to Language Structure*. Mahwah, NJ: Lawrence Erlbaum Associates. 93–111.

*Chafe, W.* (2003). "Prosodic and functional units of language". In: J. A. Edwards and M. D. Lampert (eds.), *Talking Data: Transcription and Coding in Discourse Research*. Hillsdale, NJ: Lawrence Erlbaum Associates. 33–43.

*Croft, W. and Cruse D.A.* (2004). *Cognitive Linguistics*. Cambridge: Cambridge University Press.

*Ellis, R.* (2008). *The Study of Second Language Acquisition*. Oxford: Oxford University Press.

*Hulstijn, J.* (2007). "Fundamental issues in second language acquisition". *EUROSLA Yearbook* 7, 191–203.

*Langacker, R.* (1987). *Foundations of Cognitive Grammar, Volume 1, Theoretical Prerequisites*. Stanford, California: Stanford University Press.

*Lanagacker, R.* (1991). *Foundations of Cognitive Grammar, Volume 2, Descriptive Applications*. Stanford, California: Stanford University Press

*Nęcka, E., Orzechowski, J. and Szymura, B.* (2006). *Psychologia Poznawcza* [Cognitive Psychology]. Warszawa: Wydawnictwo Naukowe PWN.

*Skehan, P.* (1998). *A Cognitive Approach to Language Learning*. Oxford: Oxford University Press.

*Slobin, D. I.* (1996). *From 'Thought to Language' to 'Thinking for Speaking*. In: J. Gumperz and S. C. Levinson (eds.), *Rethinking Linguistic Relativity*. Cambridge: Cambridge University Press. 70–96.

*Slobin, D.* (2004). "The many ways to search for a frog: linguisitc typology and the expression of motion events". In: M. Stromqvist, and L. Verhoeven (eds.), *Relating Events in Narrative*, Vol. 2: *Typological and Contextual Perspectives*. 219–257.

*Talmy, L.* (2000). *Toward a Conceptual Semantics, Vol. 1 and 2*. Cambridge, Mass.: The MIT Press.

*Zacks, J. and Tversky, B.* (2001). "Event structure in perception and conception". *Psychological Bulletin* 127, 3–21.

Daniel Karczewski
*The University of Białystok, Poland*

# Factors determining genericity in the light of experimental studies of generics

**Abstract:** The article reports on and discusses the results of three studies concerning the comprehension of generics by three different groups of speakers. It aims to discuss some of the factors (e.g., competing cues or frequency effects) that might have given rise to the conflicting interpretations. The studies seem to provide more experimental data to a growing body of evidence which supports the claim that not all learners converge on the same grammar. The experimental findings speak in favour of a well-motivated claim that plural generics represent the prototypical generic construction. The article also suggests directions for further research such as, for example, determining to what extent the differences in generic comprehension are education-related.

Keywords: generics, non-generics, genericity, competing cues.

## 1. Introduction

Generics express generalizations about members of a class. It is widely acknowledged that generics can have a variety of grammatical forms (e.g., Carlson and Pelletier 1995; Langacker 1999; Radden and Dirven 2007; Radden 2009). It is claimed that generics differ in the extent to which they express generality, as well as how many roles within a sentence are generic as opposed to specific (Langacker 1997: 194). Needless to say, there is a consensus among scholars that genericity is not a uniform phenomenon. In this connection, however, it is important to note that we distinguish between two classes of genericity: kind and characterizing predication. These two notions were earlier included under the term "genericity" (Krifka et al. 1995). A characterizing sentence (1a) allows for exceptions, by contrast, sentence (1b) which refers to a type, does not allow for exceptions. However, since the borderline between them is far from clear-cut, one is tempted to view these two distinctions as the poles of a continuum rather than in terms of a clear dividing line.

(1) a. *A lion has a bushy tail.*
    b. *The lion is a predatory cat.*

Generics can assume the following forms in English (Gelman 2003: 215):

(2) a. *The bird* is a warm-blooded animal.
b. *A cat* has 9 lives.
c. *Dinosaurs* are extinct.
d. *The elderly* need better health care.

It is important to note that what all the expressions in (2) have in common is a conceptual basis, which is to say that they refer to a kind as a whole (Gelman 2004: 445). They refer to the entire category of objects (e.g., birds, cats, dinosaurs, and elderly people as a class) and emphasize properties of the category which are stable, enduring, and timeless (Lyons 1977: 194).

The four expressions of genericity in (2) can be contrasted with non-generic expressions such as the following:

(3) a. *The bird* is sitting in the tree.
b. *A cat* caught three mice.
c. There are *dinosaurs* in the Natural History Museum in London.
d. *The elderly man* lifted the box.

The non-generic expressions in (3) highlight qualities of the category that are accidental, transient or tied to a particular context (Lyons 1977: 194).

## 2. Review of literature

This section reviews an article by Gelman and Tardif (1998) on a comparison of generic noun phrases in English and Mandarin Chinese (henceforth Mandarin), which was the main point of reference for my studies. Broadly speaking, Gelman and Tardif's (1998) studies aim to answer two general questions: (i) whether the expression of genericity which varies across languages has an impact on the generic usage and (ii) whether generics show domain differences.

To do so, a set of three studies was conducted to examine generics in two languages that express them in different ways (English relies on morphosyntactic cues, whereas Mandarin does not). Unlike English, Mandarin does not have articles, plurality or tense. Moreover, aspect is not always present in Mandarin. If we want to determine whether a sentence is generic or non-generic, we must rely on pragmatic and contextual cues. In other words, it is the absence of place markers, time or specific number that is suggestive of genericity.

Gelman and Tardif (1998: 218) pose the question of whether such linguistic differences have any implications on generic production and comprehension. One possible answer is that semantics is universal and therefore formal differences are insignificant. A number of studies, however, seem to be negating this theoretical stance by providing evidence that cross-linguistic differences in linguistic forms have an impact on semantics. Additionally, it would appear that speakers of languages in which grammatical distinctions are indispensable, 'habitually attend' more to such distinctions (Lucy 1992: 87, quoted in Gelman and Tardif 1998: 218). As far as generics are concerned, there seem to be cross-linguistic differences in their semantics (Gelman and Tardif's (1998: 218) discussion of Chierchia (ms.)).

Let us now turn to Study 1 & Study 2 and their main findings. Gelman and Tardif (1998) determined that generics were identified in Mandarin as reliably as in English. It is particularly interesting in the light of Bloom's (1981)[1] claim that Mandarin speakers do not form generic concepts and the fact that Mandarin lacks any obvious morphosyntactic cues of genericity. Moreover, Gelman and Tardif (1998) found that generics were more frequent in English than in Mandarin. The experimenters also observed that generics in both languages were the most common in the domain of animals.

Similarly to Study 1, Study 2 also confirmed that generics were readily identified in both languages. It is important to note that the overall agreement between the coders (bilingual speakers of English and Mandarin, trained in linguistics) in each language was over 90%. Moreover, the study also reveals a high production of generics in child-directed speech. The main focus of Gelman and Tardif's Study 3 was on generic comprehension. The study disclosed that genericity is a psychologically real phenomenon in both languages.

The main findings of the three studies conducted by Gelman and Tardif (1998) show that there are three major similarities between English and Mandarin, namely, generics: (i) are frequent in both languages, (ii) appear more often in book-reading contexts than in toy play contexts and (iii) are especially frequent in the domain of animals and people rather than artifacts.

## 3. Empirical methods employed in the analysis

The three studies in this paper have been inspired by Radden's observation (personal communication, July 2011) that people fail to distinguish between generic and non-generic reference.

---

1 Quoted after Gelman and Tardif (1998: 220).

## 3.1 Subjects

Participants in Study 1 included 35 native speakers of Polish, who ranged in age from 16–35 years. Study 2 comprised 37 first-year students of English at the University of Białystok, most of whom have received on average 12 years of formal instruction in English. Study 3 included 59 American speakers (32 males and 27 females) from 18 to over 60 years old of varying educational backgrounds ranging from high school to graduate degree level.

## 3.2 Materials

In Study 1, participants received a questionnaire asking for their interpretation of sentences sampled from the corpus of translation of Temple Grandin and Kathrin Johnson's book *Animals in translation* 'Zrozumieć zwierzęta', whereas Study 2 & Study 3 are based on the examples of generics and non-generics taken from Krzeszowski (1982). The sentences used were chosen so that there would be two examples of: indefinite singular generics, definite singular generics, and plural generics as well as corresponding non-generics (i.e., indefinite singular non-generic, etc.). Each participant completed a paper and pencil questionnaire (Study 1 & Study 2) or an online questionnaire using Survey Monkey, a provider of web-based survey solutions (Study 3).[2]

## 3.3 Procedure

Each subject was asked to read the sentences and judge each as referring to 'one individual member of the category', 'a few members of the category', 'most/any members of the category', or 'difficult to say'. Following each sentence were the phrases 'one', 'a few', 'most or any', and 'difficult to say' and the subjects were asked to circle the appropriate word or phrase. If the subjects selected 'one' or 'a few', it would indicate individuative reference, whereas 'most or any' would indicate generic reference.

## 3.4 Coding

Coding involved two steps. First, I evaluated sentences with the target noun phrase taking into account the above-specified factors. Then, I asked three coders trained in linguistics to check my coding. Overall agreement between the coders and me was 93%.

---

2 The online version of the survey used in Study 2 & Study 3 is available at https://www.surveymonkey.com/s/8FCYHMM.

## 4. The analysis

This section aims to discuss some of the factors that might have given rise to the conflicting interpretations and which should be neutralized before another stage of the experiment is started in the future.

### 4.1 Predication type

Let us consider one of the items used in Study 1: *Czego na przykład potrzebuje krowa w rzeźni, by być szczęśliwa?* 'What does *a cow* headed to slaughter need in order to have a happy life?' In the case of this item, ten out of thirty-five respondents marked the noun phrase *krowa* 'a cow' as an instance of individuative reference, while the coders marked it as generic. It would seem that the speakers of Polish made use of different cues in determining whether a noun phrase is generic or non-generic. The difference in marking might result from an interplay of a number of factors of which a predication type and a formal grammatical cue are most important. Thus, it would appear that the predication *szczęśliwa* 'happy' influenced the non-generic interpretation of the sentence since being happy seems to be normally predicated of individuals rather than of kinds.

### 4.2 Grammatical cue

Another factor that might be at play is the grammatical cue (i.e., the nominal in the singular), which some of the speakers might not associate with the generic reading. Generally speaking, in my Studies, the items with the plural generic nominal were more readily interpreted as generic that those which contained the nominal in the singular. For instance, in Study 1 the items with the plural nominal were almost unanimously declared as generic. More specifically, let us consider the data from Study 3. Table 1 below shows the results of the comprehension of the plural generic: 55 out of 59 respondents identified the sentence *Snakes have no legs* as generic (93,2%), while 2 subjects marked this sentence as non-generic (3,4%).

*Table 1: Response percent and response count regarding the sentence* Snakes have no legs.

| **Snakes have no legs.** | | |
|---|---|---|
| **Answer Options** | **Response Percent** | **Response Count** |
| ONE | 0.0% | 0 |
| A FEW | 3.4% | 2 |
| MOST or ANY | 93.2% | 55 |
| DIFFICULT TO SAY | 3.4% | 2 |
| | ***answered question*** | **59** |
| | ***skipped question*** | **1** |

On the other hand, there were substantial differences (as opposed to the standard coding) in the interpretation of the nominals in the singular, which ranged from 9 to 100%.

## 4.3 Frequency effects

Let us now focus on a cognitive factor which might be held responsible for some of the results from Study 3. The discussion begins with *A sparrow is a bird* since it gave rise to a variety of conflicting interpretations (Table 2 below).

*Table 2: Response percent and response count of the sentence* A sparrow is a bird

| **A sparrow is a bird.** | | |
|---|---|---|
| **Answer Options** | **Response Percent** | **Response Count** |
| ONE | 40.7% | 24 |
| A FEW | 0.0% | 0 |
| MOST or ANY | 50.8% | 30 |
| DIFFICULT TO SAY | 8.5% | 5 |
| | ***answered question*** | **59** |
| | ***skipped question*** | **1** |

Let me suggest that the surprising results regarding this item might be attributed to frequency effects. According to Evans (2007: 87), frequency, which is a key notion in a usage-based model, is claimed to have an impact on the language system. Thus, the more frequently we encounter a word or a construction, the more entrenched it becomes in the language system. I am far from suggesting that *A sparrow is a bird* is a highly entrenched construction which as a result of its entrenchment affects other less well entrenched generic construals. I am more inclined to suggest that the interpretational differences are rather due to the effect of the frequency of the type we categorize. According to Hurford et al. (2007: 89), children learn many concrete concepts such as, for example *bird*, *dog* or *man* from ostensive definitions. I assume that parents in the presence of their children frequently point in the direction of a bird in a tree saying that it is a bird and further categorizing it as for example a sparrow. Thus, the information that a sparrow is a bird is well entrenched in our minds as we have been bombarded with it since our early childhood. It is tempting to suggest that the huge incongruity between the standard coding and the way participants in the survey marked this item (to 24 out 59 subjects it is an instance of individuative reference while another 5 were unable to decide) might have been due to the item's high frequency and its entrenchment as a result. I would expect that if a different item had been chosen, for example *A penguin is a bird*, it would have returned results more in line with the standard coding.

## 4.4 Folk and expert theories

Another cognitive factor that might have given rise to different interpretations is the source of knowledge (i.e., folk and expert theories) that the information encapsulated in the generics comes from. Let us, first of all, consider two items from the survey: *An atom is composed of protons, neutrons and electrons*, as well as *The elephant is a big mammal*. This is the kind of knowledge that is most likely to be acquired in school settings. Thus, during for example a chemistry lesson, we learn what an atom is and what its components are, whereas in biology classes we might be taught that elephants are mammals because they have tiny hairs on their body and the females have mammary glands. Depending, however, on our individual experience, such information could be established in our minds with varying degrees. At the other end of the scale, we would have items such as the already discussed generic *A sparrow is a bird*, which represents knowledge that we use automatically and no conscious thought appears to be involved.

Let us now turn to the generic *The airplane uses a lot of fuel*, which is an interesting case since 14 out of 59 subjects chose the 'difficult to say' option in reference to it. It is important to emphasize that it was the only item which so many participants found difficult to classify (the level of indecisiveness in the case of the remaining generics ranged from 2 to 5).

*Table 3: Response percent and response count of the sentence* The airplane uses lots of fuel

| **The airplane uses lots of fuel.** | | |
|---|---|---|
| **Answer Options** | **Response Percent** | **Response Count** |
| ONE | 33,9% | 20 |
| A FEW | 0,0% | 0 |
| MOST or ANY | 42,4% | 25 |
| DIFFICULT TO SAY | 23,7% | 14 |
| | ***answered question*** | **59** |
| | ***skipped question*** | **1** |

I would claim that such results are due in part to various sources of knowledge which influenced the decision-making of the participants. On the one hand, it appears to be commonly known that the plane uses far more fuel than, for example the car. Thus, comparing these two means of transport, the latter is more fuel efficient than the former. On the other hand, there are some types of planes that are more fuel efficient than others such as, for example the Boeing 787 Dreamliner, which is said to use the least amount of fuel of all the airliners (according to the commercial television network TVN). The former information seems to be part of our folk knowledge of planes, whereas the latter is part of our expert theories. Thus, conflicting theories might be held responsible for interpretational differences of item *The airplane uses a lot of fuel.*

## 5. Conclusions

There are several general findings which stem from the three studies. First, the plural construal was the most reliably identified generic. This might suggest that it is the most natural expression of genericness among the three types surveyed. Furthermore, it could also indicate that the speakers had more experience with the plural generic than with the indefinite and definite constructions and hence lack a well-entrenched indefinite/definite generic schema.

Second, Study 2 & Study 3 seem to provide more experimental data to a growing body of evidence which supports the claim that not all learners converge on the same grammar (e.g., Street and Dąbrowska (2010)). To be more specific, Street and Dąbrowska (2010) present experimental evidence that native speakers differ in their native language attainment. The two experiments tested core English grammar, namely actives (e.g., *The boy photographed the girl*), passives (e.g., *The girl was hugged by the boy*), and quantificational constructions (e.g.,

*Every feather is in a vase* or *Every shoe has a hamster in it*). The data sample used in the experiment comprised actives, passives and quantificational constructions. The first experiment tested two groups of participants, namely high academic achievement (hereafter HAA) speakers and low academic achievement (hereafter LAA) speakers. The experiment showed that the HAA speakers did well on all the tasks, whereas the LAA participants did well only on actives. The second experiment sought to determine whether with some training, the LAA speakers would perform better on the tasks. Street and Dąbrowska (2010) found out that the LAA subjects performed much better after initial training in a given construction.

Third, Study 2 & Study 3 show that L2 learners outperformed native speakers as far as the comprehension of the two sentences: *A sparrow is a bird* and *The airplane uses a lot of fuel*. The study found that 50,8% of the native speakers of English understood the former sentence as generic. The figure, however, is much higher for L2 learners (70%). In the case of the latter sentence, 42,4% of American speakers and 67% L2 learners indicated it as generic. It is interesting to note that the interpretational differences between these two groups of subjects concerning the remaining generics is rather insignificant (the difference ranges from 2 to 7%). It would appear that such differences might be education-related. Polish native speakers were all university students, whereas the American participants represented different educational backgrounds (from lower than high school up to graduate degree level). It is important to note, at this point, that my initial hypothesis was that speakers with more schooling (a bachelor or graduate degree) would outperform those with less schooling (some high school) on this task. If we, however, consider the data from Table 4 in Appendix, it would be difficult to confirm this hypothesis. Let us consider the most problematic generic construal (i.e., indefinite singular) represented by sentence *A sparrow is a bird* and how it was interpreted by the three groups of participants with different educational backgrounds (i.e., some college, associate or bachelor degree and graduate degree). Thus, the item was marked as an instance of individuative reference by 10 out of 22 subjects with some college education (i.e., 45% of subjects), 6 out of 14 subjects with an associate or bachelor degree (i.e., 42% of subjects excluding 1 person who marked it as 'difficult to say') and 5 out of 12 subjects with a graduate degree (i.e., 41% of subjects excluding 3 who marked it as 'difficult to say'). Overall, I would claim that there is no conclusive evidence to suggest that the HAA speakers outperformed the LAA speakers. It is important to stress that in this study the subjects with some college education (22 out of 59) outnumbered those with a different educational background. Clearly, more research is needed to examine to what extent the differences in generic comprehension are education-related. In connection with this, it is important to report Dąbrowska and Street's (2006) findings concerning the comprehension of passive constructions by four groups of participants: (1) graduate natives, (2) graduate non-natives, (3) non-graduate natives, and (4) and non-graduate non-natives. The items used in the study comprised of sentences which were active and plausible (e.g., *The dog bit the man*), active and implausible (e.g., *The man bit the dog*), as well as passive and plausible (e.g., *The man was bitten by the dog*) and passive and implausible (e.g., *The dog was bitten by the man*). The subjects were asked to listen to the sentences and identify the agent of the sentences. One result of this study seems particularly interesting in the light of my own research into the comprehension of generics and non-generics, namely the fact that LAA non-natives performed better than LAA natives. Further research will be necessary to determine whether this result might be replicated in the case of generic sentences. Overall, Dąbrowska and Street's (2006) research seems to indicate that L2 speakers can outperform L1 speakers on some grammatical tasks.

Fourth, the results of the comprehension of sentence *A sparrow is a bird* seem rather surprising and are in need of explanation since almost half of the respondents (24 out of 59) consider the sentence to be an instance of individuative reference. This might be the effect of conceptualizing *a sparrow* as meaning each and every one or every single one (cf. the Polish expression *każdy jeden*).

Fifth, the two Studies (2 & 3) also revealed a certain shortcoming of the questionnaire as far as the comprehension of the non-generics is concerned. It would appear that the sentence *The larks in my garden are singing beautifully* was interpreted by 28,8% of the subjects as an instance of generic rather than individuative reference. Thus, the respondents selected the most or any category rather than the a few category. My hypothesis is that some of them selected the most or any option to indicate that every single bird *in my garden* sings beautifully (and not just a few of them). Thus, most/any did not indicate 'most/any members of the category' to those respondents but rather meant most/any *in my garden*.

As a closing remark, let me return to the major motivating force behind the three studies, that is Radden's observation (p.c.) that people fail to distinguish between individuative and generic reference. Having tested the questionnaires on three different groups of speakers, I am now in a position to draw a very tentative conclusion that the difference between the two types reference is not always straightforward. Thus, measuring participants' individual sensitivity to the distinction in question might constitute an interesting area of further research.

## References


*Carlson, G. N., & Pelletier, J. F. (Eds.).* (1995). *The Generic Book.* Chicago: University of Chicago Press.

*Dąbrowska, E.* (2012). Different speakers, different grammars: Individual differences in native language attainment. *Linguistic Approaches to Bilingualism* 2:3, 219–253.

*Dąbrowska, E., & Street, J.* (2006). Individual differences in language attainment: *Comprehension of passive sentences by native and non-native English speakers. Language Sciences* 28, 604–615.

*Evans, V.* (2007). *A Glossary of Cognitive Linguistics.* Edinburgh: Edinburgh University Press.

*Gelman, S. A.* (2003). *The Essential Child.* Oxford: Oxford University Press.

*Gelman, S. A.* (2004). Learning words for kinds: Generic noun phrases in acquisition. In D. *G. Hall & S. R. Waxman (Eds.), Weaving a lexicon. Cambridge, MA: MIT Press,* 445–484.

*Gelman, S. A., & Tardif, T.* (1998). A cross-linguistic comparison of generic noun phrases in English and Mandarin. *Cognition* 66, 215–248.

*Grandin, T. & Johnson, C.* (2011). *Zrozumieć zwierzęta.* Poznań: Media Rodzina.

*Hurford, J. R., Heasley, B., & Smith M. B.* (2007). *Semantics: A Coursebook.* Cambridge: Cambridge University Press.

*Karczewski, D.* (2013). *Genericity in language and thought. A cognitive study.* Unpublished Ph.D. dissertation, Institute of English Studies, Warsaw University.

*Krifka, M.* (1995). Common nouns: a contrastive analysis of Chinese and English. In G. N. Carlson & J. F. Pelletier (Eds.), *The Generic Book.* Chicago: University of Chicago Press, 398–411.

*Krzeszowski, T. P.* (1982). *Gramatyka angielska dla Polaków.* Warszawa: Państwowe Wydawnictwo Naukowe.

*Langacker, R. W.* (1999). Generic constructions. In *Grammar and Conceptualization.* Berlin and New York: Mouton de Gruyter, 247–260.

*Langacker, R. W.* (1997). Generics and habituals. In R. Dirven & A. Athanasiadou (Eds.), *On Conditionals Again.* Amsterdam and Philadelphia: John Benjamins. *Current Issues in Linguistic Theory* 143, 191–222.

*Lyons, J.* (1977). *Semantics* Vol. I. Cambridge: Cambridge University Press.

*Radden, G.* (2009). Generic reference in English: A metonymic and conceptual blending analysis. In K. Uwe-Panther, L. L. Thornburg & A. Barcelona (Eds.), *Metonymy and Metaphor in Grammar.* Amesterdam: John Benjamins Publishing Company, 199–228.

*Radden, G. & Dirven, R.* (2007). *Cognitive English Grammar*. Amsterdam: John Benjamins Publishing Company.

*Street, J. & Dąbrowska, E.* (2010). More individual differences in language attainment: How much do adult native speakers of English know about passives and quantifiers? *Lingua* 120, 2080–2094.

## Appendix

*Table 4: Response percent and response count of the sentence* A sparrow is a bird *depending on subjects' education*

| **A sparrow is a bird.** | | | | | | | |
|---|---|---|---|---|---|---|---|
| | **Education** | | | | | | |
| **Answer Options** | **Less than high school degree** | **High school degree** | **Some college** | **Associate or bachelor degree** | **Graduate degree** | **Response Percent** | **Response Count** |
| ONE | 1 | 2 | 10 | 6 | 5 | 41,4% | 24 |
| A FEW | 0 | 0 | 0 | 0 | 0 | 0,0% | 0 |
| MOST or ANY | 0 | 3 | 12 | 8 | 7 | 51,7% | 30 |
| DIFFICULT TO SAY | 0 | 0 | 0 | 1 | 3 | 6,9% | 4 |
| | | | | | | ***answered question*** | **58** |
| | | | | | | ***skipped question*** | **2** |

Ida Stria
*Adam Mickiewicz University, Poland*

# Linguistic Worldview of Esperanto: a Questionnaire Method

**Abstract:** This paper aims to identify the elements needed to carry out studies on the linguistic worldview of Esperanto.

The linguistic worldview is a subjective and socially created interpretation of reality embedded in the language. The chief objective is taking into account different levels of language that is vocabulary, idioms, morphology and syntax. Esperanto is probably the only fully developed planned language allowing for the study of its worldview.

All Esperanto speakers are at least bilingual, which leads to an assumption that the worldview in this language will contain elements transferred from other languages spoken by the respondents. Therefore, one of the research stages must contain tasks designed to detect these elements and determine the impact of the native languages.

Keywords: Esperanto, linguistic worldview, multilingualism, linguistic transfer.

## 1. The general purpose of the project

This paper aims to identify the elements needed to carry out studies on the linguistic worldview of Esperanto. The approach described in this paper, namely the questionnaire method for bilingual speakers of a planned language, has not been employed in any previous research to the knowledge of the author. Some analogies are present in the article of Koutny (2010), which is treated as a preliminary research study for the project presented here.

### 1.1 Introductory remarks on Esperanto

Esperanto is a constructed international auxiliary language created by L. L. Zamenhof, who published his pamphlet on "lingvo internacia" in 1887. His pseudonym "Doktoro Esperanto" soon began to be the name of the language itself.

Although Esperanto has been consciously and deliberately created, it is used among thousands of speakers (some sources say up to 3.5 million; see Piron, 1989), some of which are even third generation native speakers. Numerous meetings worldwide give evidence of its use in intercultural communication.

Typologically Esperanto is built in resemblance to agglutinative languages with a developed system of over 40 prefixes and suffixes, although its fundamental vocabulary is based mostly on Romance and Germanic languages. The word order is usually SVO, but free word order is available thanks to the accusative marker –n.

## 1.2 The objectives of the questionnaire approach

So far the linguistic worldview (henceforth LWV) has only been described for natural languages. Esperanto is usually considered as an artificial language, and therefore the tools of cognitive or anthropological linguistics have not been used in its description. This project is aimed at applying the theory of the LWV known from the Ethnolinguistic School of Lublin to Esperanto, which in several linguistic works has been regarded as a language developing naturally (Blanke, 1997, 2001; Koutny, 2009, 2012b; Lindstedt, 2006; Manaris, Pellicoro, Pothering, & Hodges, 2006; Schubert, 1989). Such an approach assumes the same sociolinguistic traits of both Esperanto and ethnic languages.

The scientific objective of the project is first to determine whether there is a homogeneous linguistic worldview (i.e., linguistic way of categorizing and organizing the world) in Esperanto, whose users are at least bilingual and come from different cultures. The bi- and multilingualism of the speakers leads to an assumption that the worldview in this language will contain elements transferred from their native languages. Therefore, one of the research stages must contain various tasks designed to detect these elements and determine the impact of the native languages on Esperanto. Another extremely important stage would be the study of a small group of native speakers of Esperanto and the specific worldview pertaining to the culture of Esperanto, created in the communicative practice of the Esperantist community. This problem will be taken up in detail in the following sections.

## 2. Pre-existing research

The idea of the linguistic worldview (otherwise known as the *picture of the world*) appeared for the first time in the works of W. von Humboldt in the first half of the nineteenth century, though its origins can be traced back to Herder, Luther, or even Aristotle (see Żuk, 2010). These views were developed in two different ways: by German researchers such as L. Weisgerber and by American anthropological linguists (Boas, Sapir, and Whorf). In Poland the LWV has been explored in Lublin Ethnolinguistic School (the main representative being J. Bartmiński), Wrocław (e.g., J. Anusiewicz), and Warsaw (R. Grzegorczykowa, axiolinguists with J. Puzynina).

The study of the LWV for bi- and of multilinguals as proposed in this project is a new idea. Customarily, the LWV has been studied within a single language. Comparative studies have also been postulated (cf. Bartmiński, 2009, 2012), but a study of conceptualisations in bilinguals (in combination native plus non-native or with two native languages) most likely has never been carried out in the framework of this theory. Therefore, this would seem the first attempt to describe linguistic transfer and interference within the theory of the linguistic worldview of J. Bartmiński.

Generally the problem of multilingualism has been considered from the perspective of cognitive linguistics (e.g., colours and feelings in Wierzbicka, 2004; Pavlenko, 2006; Athanasopoulos, 2009). Additionally, the concepts of interference and transfer in Esperanto have only been considered in relation to its native users (e.g., Bergen, 2001; Lindstedt, 2006).

As mentioned, the testing of the LWV in a constructed language is a new idea. A preliminary study was conducted by Koutny (2010). The analysis was based on a questionnaire containing five questions on personal details and nine complex questions concerning the linguistic worldview. The questions related to both the grammar and the vocabulary of Esperanto. 100 speakers of 19 languages responded, of which four persons declared themselves as native speakers of Esperanto. The study has shown that Esperanto users follow their native languages when naming the colour of objects or assessing to which class (e.g., animate/inanimate) a notion belongs. However, some notions are understandable only to those familiar with Esperanto culture and actively participating in the community. Synthetic morphological forms appear more often, regardless of the type of native language.

## 3. The method

### 3.1 Preliminary requirements

In this project the theory will be adopted from the works of J. Bartmiński, although the research method will not be identical to the proposals of Lublin School and the methodology adopted in the EUROJOS project. The data required to describe the LWV according to Bartmiński (2009), Grzegorczykowa (1999) and in EUROJOS are as follows: (a) systemic (including hyponyms, opposites, synonyms, derivatives, collocations and proverbs), (b) dictionary definitions, and (c) real-life instances from texts, corpora and questionnaires. However, the questionnaires need only have one question, namely "what is a true X like?" while additional questions might, but do not have to be asked.

Although Esperanto is treated here as natural and meets all the theoretical prerequisites postulated by the EUROJOS team (i.e., has original literature, corpora, several dictionaries, thousands of speakers and some dozens of native speakers), there are technical constraints to the task. Any study of Esperanto would necessitate a large number of team members and would last several years. Therefore, a basic introductory study is needed before more detailed subsequent projects are conducted precisely in accordance with Bartmiński's method. Such an introductory questionnaire study is presented below.

## 3.2 Basic assumptions

The linguistic worldview according to a concise definition of Jerzy Bartmiński (2009) is:

> a language-entrenched interpretation of reality, which can be expressed in the form of judgments about the world.... The interpretation is a result of subjective perception and conceptualization of reality performed by the speakers of a given language; thus, it is clearly subjective and anthropocentric but also intersubjective (social). (p. 23)

This definition refers not only to the interpretation generated in a community (a condition that is satisfied by Esperanto), but also to individual conceptualisations derived from experience. As mentioned, all users of Esperanto are at least bilingual, which could mean that the worldview embedded in this language is mediated by the speakers' native languages.

The results of the study of Koutny (2010; see section 2), together with the theoretical analysis carried out in Stria (2013), led to the formulation of the following research questions:

Q1: Will the LWV of non-native users be taken from their native language (L1)?
Q2: Is there a homogeneous, culturally embedded LWV of Esperanto, understandable for non-native speakers?
Q3: Is there a homogeneous LWV among native speakers of Esperanto?

Answers to these questions are to be obtained in a procedure comprising the following research tasks: (a) establishing a list of survey questions, (b) preparing questionnaires and conducting surveys among respondents, (c) choosing and grouping questionnaires according to established parameters, (d) comparative analysis of questionnaires filled out by non-native respondents, (e) analysis of questionnaires filled out by native speakers.

This project will cover a different group of respondents than those of Koutny (2010). The study sample will include approx. 100 advanced users of Esperanto. The data is to be collected by selecting four equal groups of native speakers of different languages distant from each other in order to statistically compare the results. The first three groups of 25 persons each are to be as follows:

*Table 1: Samples according to language*

| culture / language | European | Non-European |
|---|---|---|
| Indo-European | e.g., Polish, Swedish, German | ------ |
| Non-Indo-European | Hungarian | e.g., Chinese |

It is visible from the table that the combination non-European culture plus Indo-European language (e.g., Hindi) is missing. The reason for this is the limitations of resources. Ideally, all four groups should be used. The choice of the languages should depend on the knowledge of those languages of the researchers.

The fourth group included in the project will be at least 20 (ideally 25) native speakers of Esperanto. The diversification of the research pool according to language spoken will allow for determining the impact of language and culture on categorisations. It seems that 25 respondents from each of the groups is a number large enough to control for the idiosyncrasy, while small enough to successfully carry out the project within the stipulated time (up to 24 months).

The questions will be written exclusively in Esperanto to prevent transfer of structures from another language on the responses in the questionnaires. The responses (where possible) will be compared with LWVs known from L1s of the users (this procedure has not been used by Koutny), which will determine the scope of interference and transferring from L1. Subsequent studies should also include translation tasks designed to determine the extent of transferring.

## 3.3 The tasks

The first stage of the project is establishing a list of survey questions. The questionnaire will consist of an introductory part with questions on personal details, taking into account sex, age, native language, and the command of Esperanto (e.g., how often do you use Esperanto?). The remaining part of the questionnaire will only be accessible to the advanced users (of C1 level at least, native speakers included). The rest of the questions will relate to the linguistic worldview of the respondents.

The questionnaires will be distributed via Internet through national Esperanto associations, mailing lists and other media, for example social networking sites. In order to diversify the data it is planned to send out the questionnaires to users of both Indo-European and non-Indo-European languages.

After the data has been collected (until the minimum of 25 questionnaires of each of three languages are collected with at least 20 native speakers in group four) there will be a brief period of quantitative analysis, that is statistical categorisation of language, sex and age as well as coding the responses, that is grouping them by descriptors. The answers relating to syntax and morphology will be segmented into basic sentence parts and morphemes, and compared with grammatical schemes occurring in languages of the control group (i.e., three unrelated languages of the greatest number of respondents). For example, the sentence *Mi devus fari ĝin* [I should do it] is coded as S – Aux – V – O and compared with the Polish scheme (S) – Aux – O – V *Powinienem to zrobić*. The answers concerning vocabulary, collocations and idioms will be coded according to domains and descriptors. The LWVs of the groups will be statistically compared in task 4.

After the division, the questionnaires filled in by non-native speakers of Esperanto will be analysed qualitatively. If most responses are matching (i.e., mutually compatible, or compatible in terms of descriptors), it will be assumed that Esperantists of different languages and cultures share a coherent and consistent LWV (which might confirm the strength of Esperanto culture and a weak version of the Sapir-Whorf hypothesis). In the opposite case, that is the lack of agreement, the responses grouped according to the respondents' native language will be compared with worldviews known from those L1s in order to determine whether there is any conceptual interference and transfer from native languages to Esperanto. Those parts that do not have their counterparts in the native languages will be considered as belonging to the worldview of Esperanto. This task has two stages: S1, where only strictly cultural concepts (e.g., the importance of green, the term "citizen" and the five-pointed star) are examined, as well as S2, where cross-linguistically varying domains are compared (e.g., colours, categorisation of animals and plants and the definitional characteristics of astronomical objects).

The analysis of the questionnaires completed by native speakers will be carried out in the same manner. If the worldviews of these respondents will not be uniform, individual responses will be compared with the second native language of the respondents who, as previously stated, are at least bilingual.

### 3.4 Sample questions

The first part of the questionnaire will be questions on personal details, for example about age, sex, native language and command of Esperanto. Sample questions are as follows (cf. Li, Sepanski and Zhao, 2006):

1. Elektinte legi tekston disponeblan en ĉiuj lingvoj konataj al vi, en kiom da procentoj de kazoj vi elektus legi ĝin en iu de viaj lingvoj? Supozu, ke la originalo

estis skribita en alia lingvo, kiu estas nekonata al vi. [When choosing to read a text available in all languages known to you, in what percentage of cases would you choose to read it in each of your languages? Assume that the original was written in another language, which is unknown to you.]

2. En kiuj lingvoj vi kutime [In which languages do you usually]:
   - Kalkulas [Count and do simple arithmetic]?
   - Sonĝas [Dream]?
   - Esprimas sentojn [Express feelings]?
   - Preĝas [Pray]?
   - Malbenas [Swear]?

The second part of the questionnaire will only be accessible to the advanced users (C1 level at least, native speakers included). The chief objective in the selection of questions will be taking into account different levels of language, that is vocabulary, idioms, morphology and syntax as proposed by Anusiewicz, Dąbrowska and Fleischer (2000), Bartmiński (2009, 2012), Grzegorczykowa (1999). To establish the list of questions the assumptions of Rosch (1978; as an addition to the original Lublin LWV theory), the EUROJOS project, as well as the preliminary study of Koutny (2010) will also be used. Therefore, the questionnaire will contain open-ended questions: "complete", "name X", definitional, the BUT test, sentence transformation, and so forth.

In the case of heterogeneity of the worldview the answers will be compared with the already known worldviews of the native languages of the respondents in order to capture elements transferred from L1 to Esperanto. Those parts that are not found in the native languages will be considered as belonging to the worldview of Esperanto.

Sample questions will be as follows:

1. Kian koloron havas? [what is the colour of:]
   - Suno [the sun]
   - Sablo [sand]
   - Papriko [bell pepper]
   - Vulpo [fox]
   - Ĉielo [the sky]

2. Priskribu mallonge (se eblas, per unu – du vortoj) [Describe briefly (if possible, use one – two words)]:

- Li ĉiam estis aktiva esperantisto, sed ĵus forlasis la movadon. (Kion li faris?) [He was always an active Esperantist, but has just left the movement. (What has he done?)]
- Ili renkontiĝas nur dum kongresoj. Tio estas efemera amafero. (Kio ŝi estas?) [They meet only during congresses. This is a short-lived love affair. (What is she?)]
- Kiam ili estas inter esperantistoj, ili ofte parolas en sia denaska lingvo. (Kion ili faras?) [When they are among Esperanto speakers, they often speak in their native language. (What do they do?)]

The first question is designed to test whether Esperanto assigns a specific colour to each term, or whether these colours are borrowed from native languages. The second question verifies knowledge of Esperanto culture and idioms related to it. Other questions include: filling in the blanks (collocation test), questions about personifications (whether the sun is male, female, a child, etc.), prototypicality of plants and animals (e.g., "list five birds, plants, vegetables"), grammatical transformations (semantic compositionality test; see Koutny, 2012a, p. 119), the BUT test ("Complete the sentence: 'John is an ardent Esperantist, but…'"). The data will be coded according to descriptors (e.g., coral, crimson, burgundy – descriptor: red) and subjected to statistical analysis. This coding can be explained on the example of the question: "According to you, what is the real Esperantist like?" Responses will be grouped into domains (e.g., social aspect/ ideological aspect/ physical aspect, etc.) in which the keywords will be placed (e.g., respect, altruism/ equality/ green) extracted on the basis of the responses (e.g., *respects others and helps them selflessly/ believes that all are equal/ dresses in green*).

## 4. Conclusions

The results of the project will be of both theoretical and practical importance. Firstly, the theory of LWV will be applied to a constructed language for the first time. This will improve the understanding of the development of languages in general and the ways of conceptualizing the world using a language. Moreover, the weak version of the Sapir-Whorf hypothesis will be tested.

Secondly, the results will also contribute to the understanding of how native users of different languages speaking the same language in certain social situations produce new culturally embedded meanings.

The last research question (Q3) will concern Esperanto native speakers (do they present a consistent LWV). This question posits a fundamental linguistic problem. Not only are native Esperanto speakers never monolingual and therefore subject to the influence of another language (or several other languages), but also do not remain in close and constant contact with the speech community, which allows for questioning the consistency of their LWV. What is more, many a time Esperanto is not the dominant language, and with the passage of time it may even decline. Note that native speakers never set the standards of Esperanto, which means they do not fulfil the same function as native speakers of (and in) other languages (cf. Fiedler, 2012). A negative answer to the Q3 question could explain the ways of formation of the linguistic categorisations in multilingual users staying in non-Esperanto environment, while a positive one might indicate the validity of the weak version of the Sapir-Whorf hypothesis.

It is significant that this project can be a contribution to a further detailed study of Esperanto and a pioneering research on other constructed languages (e.g., Ido, Interlingua).

## References

*Anusiewicz J., Dąbrowska A., & Fleischer M.* (2000). Językowy obraz świata i kultura. Projekt koncepcji badawczej [Linguistic Worldview and Culture. A Research Project]. In J. Anusiewicz & A. Dąbrowska (Eds.), *Język a Kultura* 13. *Językowy obraz świata i kultura* (pp. 11–44). Wrocław, Poland: Wydawnictwo UWr.

*Athanasopoulos, P.* (2009). Cognitive Representation of Colour in Bilinguals: The Case of Greek Blues. In *Bilingualism: Language and Cognition*, 12, 83–95.

*Bartmiński, J.* (2009). *Aspects of Cognitive Ethnolinguistics*. London: Equinox Publishing Ltd.

*Bartmiński, J.* (2012). *Językowe podstawy obrazu świata* [Linguistic Foundations of the Worldview] (4th ed.). Lublin, Poland: Wydawnictwo UMCS.

*Bergen, B.* (2001). Nativization Processes in L1 Esperanto. In *Journal of Child Language* 28(3). 575–595.

*Blanke, D.* (1997). The Term "Planned Language". In H. Tonkin (Ed.), *Esperanto, Interlinguistics, and Planned Language* (pp. 1–20). Lanham, MD: University Press of America.

*Blanke, D.* (2001). Vom Entwurf zur Sprache [From Draft to Language]. In K. Schubert, (Ed.), *Planned Languages: From Concept to Reality* (pp. 37–89). Brussels, Belgium: VLEKHO.

EUROJOS. (2009). Retrieved September 15, 2014 from the Institute of Slavic Studies, Polish Academy of Sciences at http://ispan.waw.pl/default/en/research-projects/eurojos.

*Grzegorczykowa, R.* (1999). Pojęcie językowego obrazu świata [The Notion of the Linguistic Worldview]. In J. Bartmiński (Ed.), *Językowy obraz świata*. Lublin, Poland: Wydawnictwo UMCS.

*Fiedler, S.* (2012). The Esperanto Denaskulo. The Status of the Native Speaker of Esperanto within and beyond the Planned Language Community. In *Language Problems & Language Planning*, 36/1, 69–84.

*Herring, J.* (2005). *Syntactic and Lexical Changes in Esperanto: a Corpus-based Survey*. 2nd Midwest Computational Linguistics Colloquium, Columbus, OH. Retrieved September 15, 2014 from: http://cllt.osu.edu/mclc/paper/syntactic_herrring.pdf.

*Koutny, I.* (2009). Esperanto im Rahmen der Sprachtypologie [Esperanto in Frames of the Language Typology]. In S. Fiedler (Ed.), *Esperanto und andere Sprachen im Vergleich* (pp. 117–130). Berlin: Gesellschaft für Interlinguistik.

*Koutny, I.* (2010). Esperantlingva bildo de la mondo [Esperantic Linguistic Picture of the World]. In D. Blanke & U. Lins (Eds.), *La arto labori kune. Festlibro por Humphrey Tonkin* (pp. 290–305). Rotterdam, the Netherlands: UEA.

*Koutny, I.* (2012a). From semantic networks to dictionary structures. In *Język. Komunikacja. Informacja*, VII, 115–128.

*Koutny, I.* (2012b). Kiel niaj pensoj vortigxas? Kiun semantikan modelon sekvas esperanto? [How Are Our Thoughts Verbalised? Which Semantic Model Does Esperanto Follow?] In Ch. Kiselman & M. Maradan (Eds.), *Leksikologio, frazeologio, historio, semantiko kaj terminologio: du kontinentoj renkontiĝas en Hanojo* (pp. 35–48). Rotterdam, the Netherlands: UEA.

*Li P., Sepanski S., & Zhao X.* (2006). Language History Questionnaire: A Web-based Interface for Bilingual Research. In *Behavior research methods*, 38(2), 202–210.

*Lindstedt, J.* (2006). Native Esperanto as a Test Case for Natural Language. In M. Suominen et al. (Eds.), *A man of measure: Festschrift in honour of Fred Karlsson on his 60th birthday* (pp. 47–55). Turku, Finland: Linguistic Association of Finland.

*Manaris B., Pellicoro L., Pothering G., & Hodges H.* (2006). Investigating Esperanto's Statistical Proportions Relative to Other Languages Using Neural Networks and Zipf's Law. In Proceedings of the 24th IASTED International Conference on Artificial Intelligence and Applications (pp. 102–108). Innsbruck, Austria: ACTA Press.

*Pavlenko, A. (Ed.).* (2006). *Bilingual Minds: Emotional Experience, Expression, and Representation*. Clevedon, UK, New York, Toronto: Multilingual Matters.

*Piron, C.* (1989). Who Are the Speakers of Esperanto? In K. Schubert (Ed.), *Interlinguistics: Aspects of the Science of Planned Languages* (pp. 157–172). Berlin, Germany: Walter de Gruyter.

*Stria, I.* (2013). Classifications of Artificial Languages. In *Język. Komunikacja. Informacja*, VIII, 125–132.

*Schubert, K.* (1989). Interlinguistics – Its Aims, Its Achievements, and Its Place in Language Science. In K. Schubert (Ed.), *Interlinguistics: Aspects of the Science of Planned Languages* (pp. 7-44). Berlin, Germany: Walter de Gruyter.

*Wierzbicka, A.* (2004). Preface: Bilingual lives, bilingual experience. In *Journal of multilingual and multicultural development*, 25 (2–3), 94–104.

*Żuk, G.* (2010). Językowy obraz świata w polskiej lingwistyce przełomu wieków [Linguistic Worldview in Polish Linguistics at the Turn of the Centuries]. In M. Karwatowska & A. Siwiec (Eds.), *Przeobrażenia w języku i komunikacji medialnej na przełomie XX i XXI wieku* (pp. 239–257). Lublin, Poland: Wydawnictwo Drukarnia Best Print.

Krzysztof Kosecki[1]
*University of Łódź, Poland*

# Comprehension of metaphor-based non-literality in signed languages by the hearing persons

**Abstract:** The ability of the hearing persons to understand the form-meaning relation of non-literal signs may be a factor contributing to mastering signed communication (Pizzuto, Boyes-Braem, & Volterra 1996; Grote & Linz 2003). In an anonymous experiment aimed to test it, the pupils of two Łódź grammar-schools were asked to comment on the relation between form and meaning of metaphor-based signs from three unrelated languages: American, British, and Polish. Each of the seven stimulus-based tasks involved signs based on various conceptual metaphors. The results showed the understanding of the motivation for the form-meaning continuum to be dependent upon sign iconicity, knowledge of the meanings of gestures, the complexity of the conceptual metaphors involved, and age.

Keywords: gesture, iconicity, metaphor, motivation, signed language.

## 1. Introduction

On August 19th 2011, the Polish Parliament passed the Bill on Sign Language and Other Means of Communication in order to implement the EU regulations into the Polish legal system. The law came into force on April 1st 2012. Its major purpose was to facilitate the normal functioning of the deaf people in the society by training sign language translators and increasing the availability of sign language courses.

The term *Sign Language* means *Signed Polish* (*System Językowo-Migowy/SJM*) – a system taught for public purposes, different from Polish Sign Language (*Polski Język Migowy/PJM*). *SJM* has many structural affinities with phonic Polish, but shares many signs with *PJM*.

1 Marcin Trojszczak, MA, is the author of percentage rate calculations and diagrams representing them.

## 2. Signed languages: non-literality

In various languages, many non-literal signs are based on metaphor and metonymy (Brennan, 1990; Grzymska, 2008; Sutton-Spence & Woll, 2010; Wilcox, 2000; Wilcox, 2008). Such signs involve cognitive iconicity, which is defined as "a relationship between our mental models of image and referent" (Taub, 2001, p. 19). Because such models rely both on universal and culture-specific embodied experience (Taub, 2001, pp. 19–20), the ability to recognize and interpret them may help the potential learners to master signed communication.

Analyses of signed metaphors show them to be as varied and complex as those functioning in phonic languages. Wilbur (1987), P. P. Wilcox (2000), S. Wilcox (2008), and Taub (2001) give examples of metaphor-based signs in American Sign Language/ASL. Equally diverse and complex metaphors are present in British Sign Language/BSL (Brennan, 1990; Sutton-Spence & Woll, 2010) and *PJM* (Grzymska, 2008).

## 3. The hypothesis

The hypothesis to be tested by the anonymous experiment was that the hearing persons should to a large extent be able to understand non-literal metaphor-based signs. That is because the signs involve the same metaphors that they use to comprehend the relevant concepts. Furthermore, in contrast to the deaf, the hearing persons' communicative capability is holistic on cognitive, social, and emotional levels: it involves phonic language and gestures, and there is no sensory barrier to effective reception of sound data and visual stimuli (Podgórska-Jachnik, 2004, pp. 66–73).[2]

## 4. The experiment

The participants – 93 hearing pupils of Łódź grammar-schools – were divided into two groups: 73, aged 14–16, attended the first level (Pol. *gimnazjum*); 20, aged 18–19, attended the second level (Pol. *liceum ogólnokształcące*). The task-based part was preceded by a short questionnaire focusing on the participants' knowledge of signed communication.

---

2 See Białas (2007) for the description of the reduced ability to comprehend metaphors in written texts by the Polish deaf.

## 4.1 The questionnaire

Please answer the following questions by marking the selected point.

1. Who in your family is a deaf or hearing-impaired person? You can mark more than one point, e.g. *c* and *d*.
   a/ No-one. b/ Parents. c/ One parent. d/ Brother or sister. e/ Grandfather or grandmother.
2. How do you communicate with other people?
   a/ Only by means of speech and writing.
   b/ By means of speech, writing, and sign language.
3. How long have you been able to communicate with others by means of sign language? Answer the question only if it applies and give the number of years.
4. If you communicate by means of sign language, how did you learn it?

The purpose of the questionnaire was to learn whether the participants had been in any way exposed to signed communication. The knowledge of *PJM* or *SJM*[3] are factors that might influence the answers to the task-based part.

Following the results of the questionnaire, the 14–16-year-olds were divided into two sub-groups: A – 64 persons having no previous contact with signed communication; B – 9 persons having some knowledge of *SJM*. No 18–19-year-olds reported any previous contact with signed communication.

## 4.2 The task-based part

Seven stimulus-based tasks involved signs from three unrelated languages – ASL, BSL, and *PJM*. Four tasks asked the participants to give motivation for the form-meaning continuum of signs based on various conceptual metaphors. In two cases, pairs of signs came from two different languages, that is, ASL and BSL; two tasks were based on single *PJM* signs. The remaining three tasks asked the participants to match one of a pair of *PJM* signs with the given concepts and justify their choice; to ensure maximum objectivity, signs that did not allow for explicit interpretation were used. The results, only for the absolutely correct responses, were estimated in terms of number and the related percentage rates.

1. The ASL and BSL signs below express the concept of 'help'. Why, in your opinion, do they involve such configuration of hands?

3 Hearing children of deaf parents (Eng. CODA – Child of Deaf Adult) can be fluent users of phonic Polish and *PJM*. Some of them may, however, opt for *SJM* as closer to the spoken language patterns (Czajkowska-Kisil, 2008, pp. 13–14).

*Figures 1 and 2: The ASL and BSL signs for 'help' (Lane, 1990, p. 82; Smith, 2010, p. 61).*

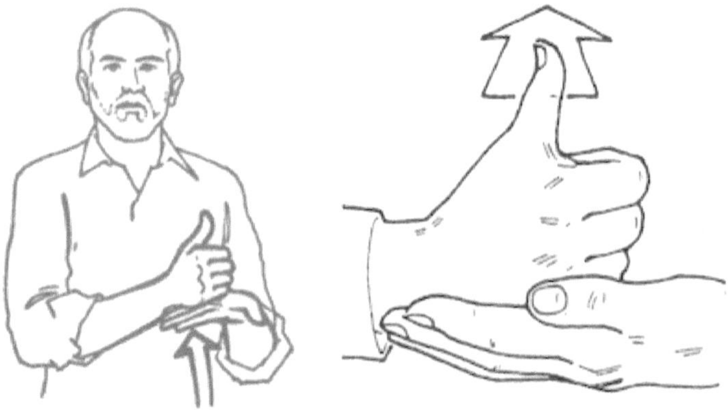

In sub-group A of the 14–16-year-olds, 5 out of 64 (7.81%) participants correctly interpreted the signs as expressing the idea of physical support, which is the source domain of the primary metaphor Help Is Support (Grady 1997, as cited in Lakoff & Johnson, 1999, p. 52) underlying them. 6 wrong interpretations said the thumb-up element meant 'OK' because it is good to help others; 4 said that the thumb-up element meant 'calling for help'; 3 said the signs expressed the sense of connection between people helping each other. The remaining 46 participants did not interpret the signs at all. In sub-group B, 0 out of 9 (0%) participants interpreted the sign correctly. 2 wrong interpretations said that the thumb-up element meant 'OK', and 1 said that it meant 'calling for help'.

*Figures 3 and 4: 'Help' – sub-groups A and B of the 14–16-year-olds.*

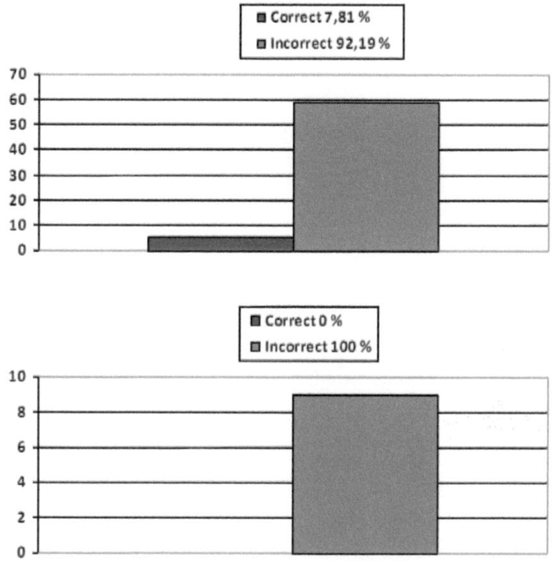


The knowledge of the common thumb-up gesture influenced the interpretation of the signs' meaning; however, only 8 participants, 6:2 in the respective sub-groups, pointed out its relation to the positive axiological aspect of helping other people.

11 out of 20 (55%) of the 18–19-year-olds correctly interpreted the signs as expressing the idea of physical support. 2 interpretations, including 1 wrong, said the thumb-up element meant that helping people was good. 1 wrong interpretation said that the signs expressed closeness of people helping each other. 9 interpretations, including 6 correct, also said the thumb-up element represented a person in need of help.

*Figure 5: 'Help' – the 18–19-year-olds.*

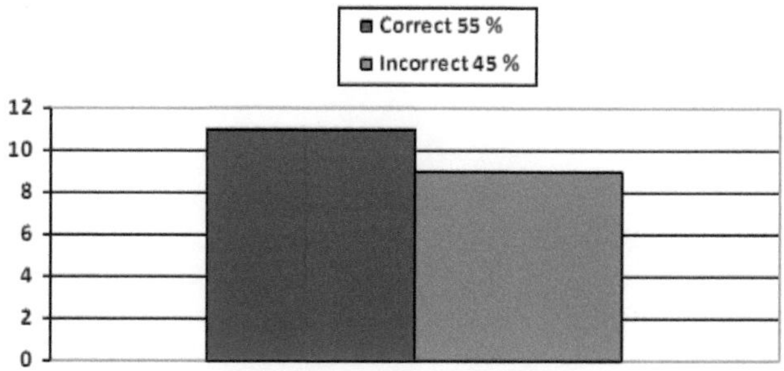

The wrong interpretations, though less numerous, mentioned the same associations as their counterparts in the younger age group.

2. The ASL and BSL signs below mean 'to agree'. Which element is, in your opinion, most closely related to the meaning and why?

*Figures 6 and 7: The ASL and BSL signs 'to agree' (Lane, 1990, p. 5; Smith, 2010, p. 128).*

In sub-group A of the 14–16 year-olds, 2 out of 64 (3.13 %) participants correctly pointed out the element of proximity, which is the source domain of the complex metaphor Agreement is Closeness underlying both signs. 2 participants gave indirectly correct interpretations related only to the BSL sign, which signified two persons being close to each other. 1 partly correct interpretation said that the ASL sign expressed the idea of understanding as a part of agreement because one of the articulators was located close to the head; 34 said the thumb-up element of the BSL sign signified the positive axiological aspect of agreement. 25 participants did not interpret the sign at all. In sub-group B, 0 out of 9 (0 %) participants correctly interpreted the signs. 5 wrong interpretations said the thumb-up element of the BSL sign conveyed positive axiological aspect of agreement.

*Figures 8 and 9: 'To agree' – sub-groups A and B of the 14–16-year-olds.*

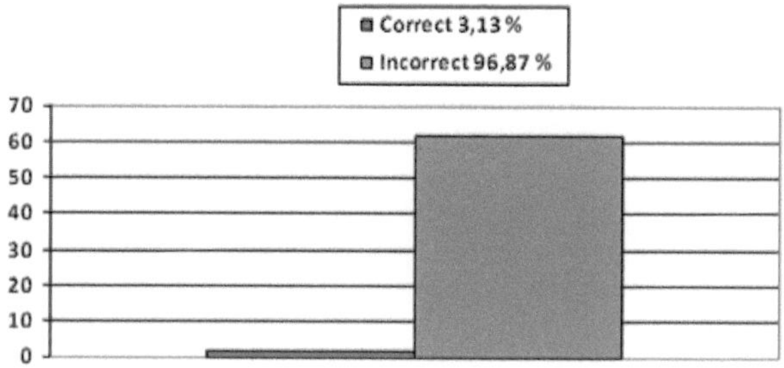


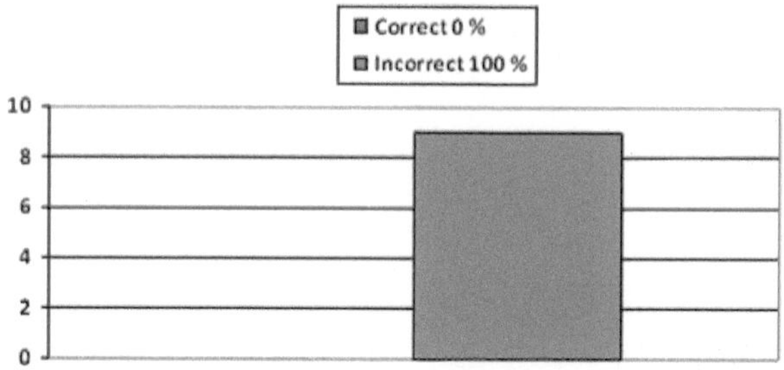


On the whole, very few fully correct interpretations were given. For the second time, influence of the common thumb-up gesture upon the interpretation of the signs was observed.

11 out of 20 (55%) of the 18–19-year-olds correctly interpreted the element of proximity, but 3 also pointed out the thumb-up axiological aspect of the BSL sign. 6 wrong interpretations mentioned the thumb-up axiological element only. 1 participant did not interpret the sign.

*Figure 10: 'To agree' – the 18–19-year-olds.*

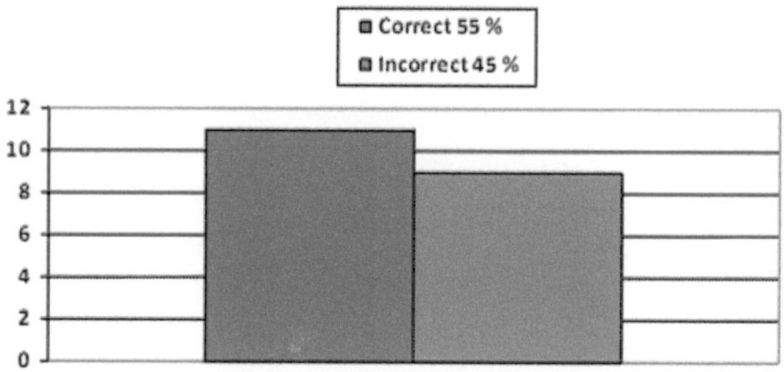

For the second time, more than a half of interpretations were correct. A part of correct interpretations, as well as many of the wrong ones, reflected strong influence of the thumb-up gesture.

3. The *PJM* sign below expresses the concept of 'court of justice'. Do you think the hand movements are in any way related to its meaning?

*Figure 11: The* PJM *sign for 'court of justice' (ESLC, 2014).*

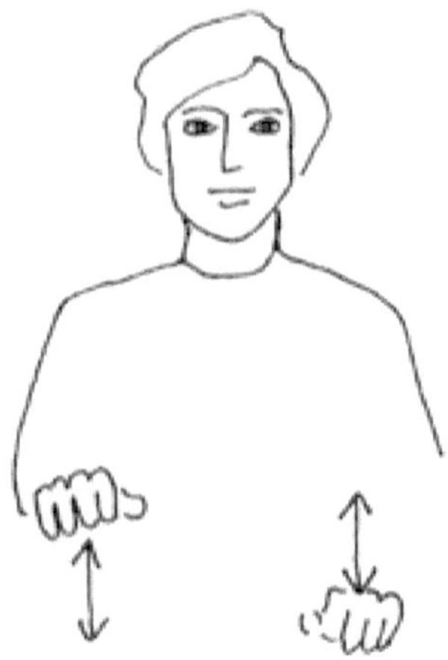

In sub-group A of the 14–16-year-olds, 14 out of 64 (21.87 %) participants correctly interpreted the sign as expressing the action of weighing, which is the source domain of the complex metaphor Judging Is Weighing. 3 wrong interpretations mentioned the judge's use of the mallet; 1 mentioned a choice out of two options; 1 pointed out the aspect of the degree to which the facts evaluated by the judge could be good or bad; 1 mentioned the judge's indecision as a part of passing the sentence. The remaining 44 participants did not interpret the sign. In sub-group B, 3 out of 9 (33.33 %) participants correctly interpreted the sign as expressing the metaphor of weighing.

*Figures 12 and 13: 'Court of justice' – sub-groups A and B of the 14–16-year-olds.*

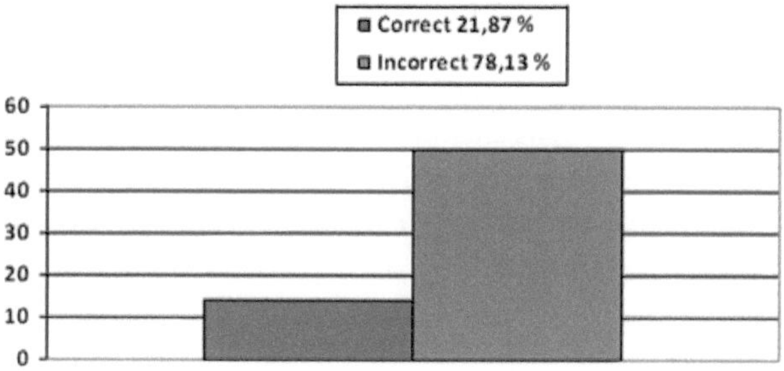


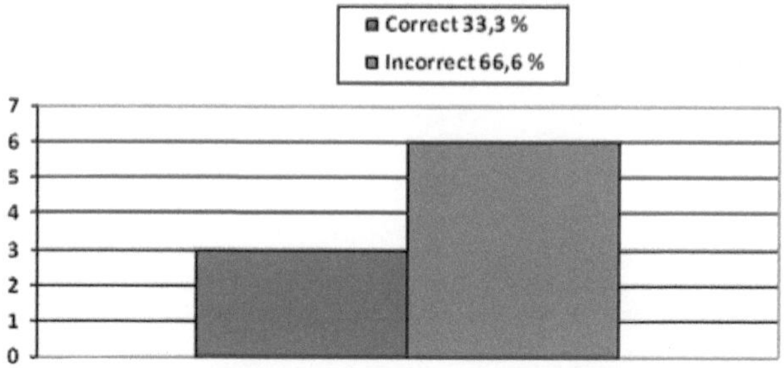


Compared to the previous tasks, more correct interpretations were provided in both sub-groups.

14 out of 20 (70 %) of the 18–19-year-olds gave correct interpretations related to the concept of weighing. 2 wrong interpretations mentioned the judge's behaviour; 1 mentioned the axiological aspect of the up down orientation.

*Figure 14: 'Court of justice' – the 18–19-year-olds.*

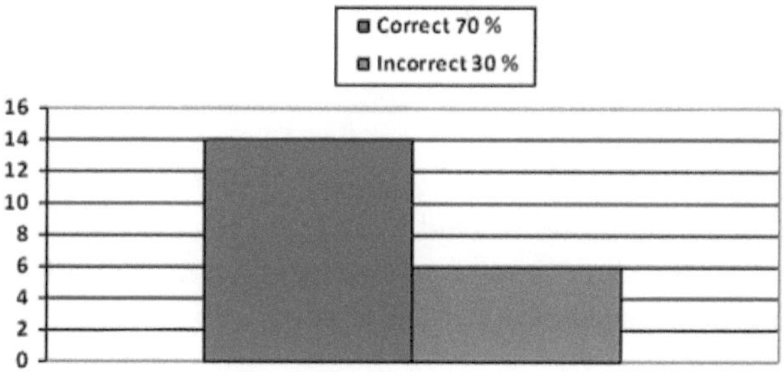

For the third time, more than 50 % of the interpretations were correct. 4 participants mentioned the mythology-related element of the Greek goddess Temida holding the scales of justice, which can be attributed to the group's educational profile including the classical subjects. The results may thus to some extent reflect Landauer and Dumais' (1997, as cited in Ritchie, 2013, p. 83) view that sources of some metaphors are in written texts rather than in direct experience.

4. The *PJM* sign below expresses the concept 'to think'. Why, in your view, the hands make circular movements?

*Figure 15: The* PJM *sign 'to think' (Hendzel,* 1995, n. pag.)

In sub-group A of the 14–16-year-olds, 12 out 64 (18.75 %) participants gave correct interpretations related to the idea of machine wheels turning, which is the source domain of the complex metaphor The Mind Is a Machine (Lakoff & Johnson, 1980, pp. 27–28) underlying the sign. 13 partly correct interpretations mentioned only the location close to the head. 39 participants did not respond to the task at all. In sub-group B, 3 out 9 (33.33 %) participants gave correct interpretations. 2 partly correct interpretations mentioned the location close to the head. 4 pupils did not respond to the task at all.

*Figures 16 and 17: 'To think' – sub-groups A and B of the 14–16-year-olds.*

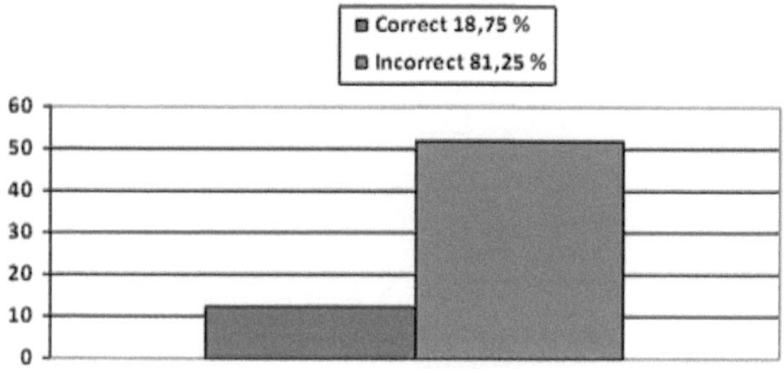


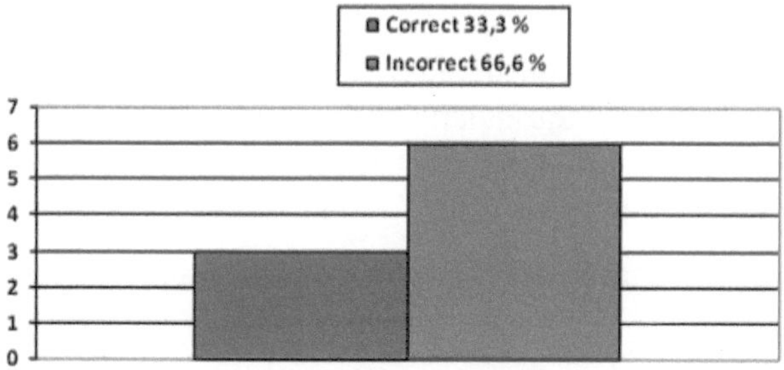


For the third time, a conventional gesture strongly influenced the interpretations: the folk model of the head being the locus of mental processes is present in both phonic and signed languages.

5 out of 20 (25 %) of the 18–19-year-olds provided correct machine-related interpretations of the sign; however, only 1 mentioned the machine explicitly, and 4 referred to work and motion. 12 wrong interpretations mentioned thoughts 'moving' in the head; 1 mentioned the brain.

*Figure 18: 'To think' – the 18–19-year-olds.*

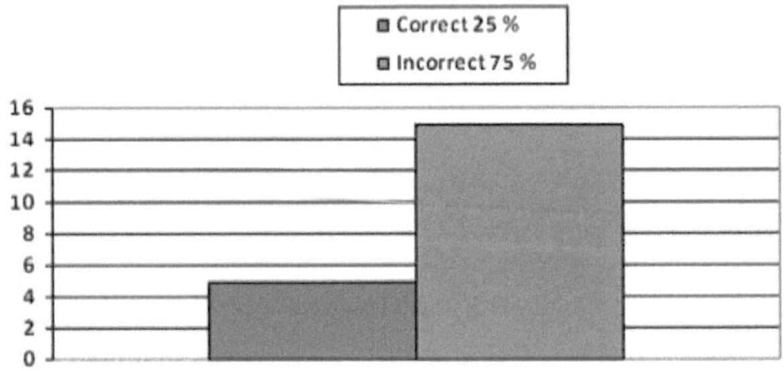


Small number of correct interpretations compared to the previous three signs may be due to the complexity of the underlying metaphor. More answers were related to a different metaphor of thinking, which means that various persons understand the same signs differently.[4]

5. Which of the *PJM* signs below expresses the concept of 'crisis' and why?

*Figures 19 and 20: The* PJM *signs for 'crisis' and 'alliance' (ESLC, 2014).*

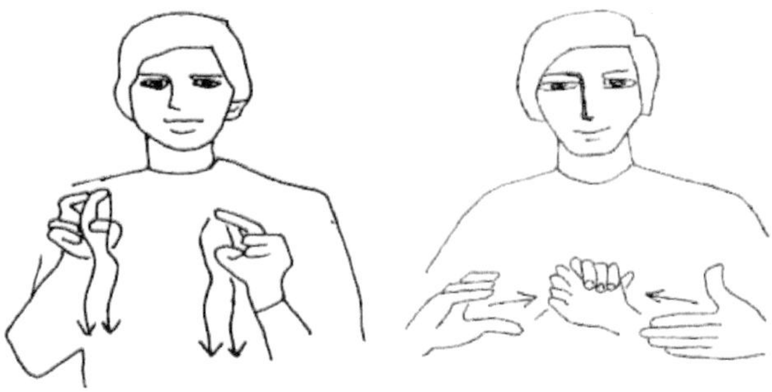

In sub-group A of the 14–16 year-olds, 20 out 64 (31.25 %) participants correctly pointed out and interpreted the sign, mentioning the movement down as a vital component of its meaning. 14 participants correctly pointed out the sign, but gave no interpretations. 30 pupils did not respond to the task at all. In sub-group B, 3 out 9 (33.33 %) participants correctly pointed out the sign and interpreted the

4 See Ritchie (2013, p. 82) for a similar claim related to analysis of linguistic expressions.

metaphorical movement down. 2 participants correctly pointed out the sign, but gave no interpretations. In 4 cases, there was no response to the task.

*Figures 21 and 22: 'Crisis' – sub-groups A and B of the 14–16-year-olds.*

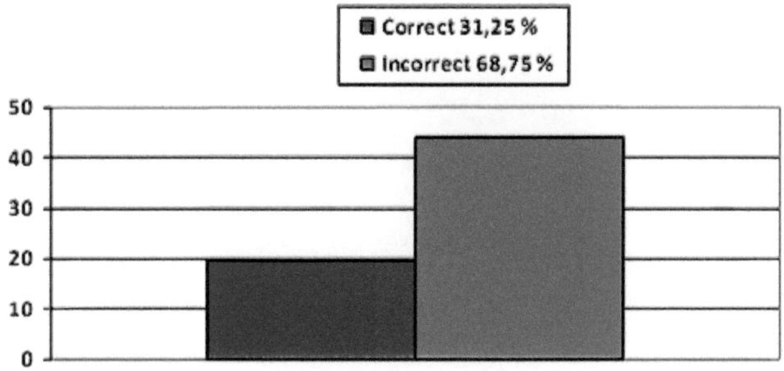


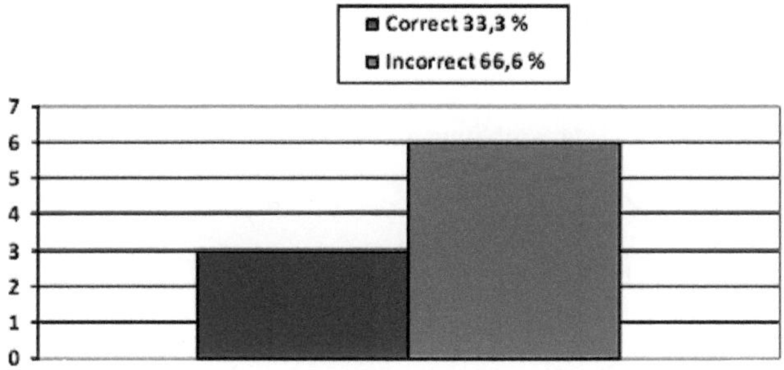


A significant number of correct indications and interpretations in both sub-groups may be due to the fundamental importance of the up-down orientation being the source domain of the primary metaphor More Is Up; Less Is Down (Lakoff & Johnson, 1980, pp. 15–16; 1999, p. 51) underlying the sign.

15 out of 20 (75 %) the 18–19 year-olds correctly pointed out and interpreted the sign. 5 participants only pointed out the correct sign.

*Figure 23: 'Crisis' – the 18–19-year-olds.*

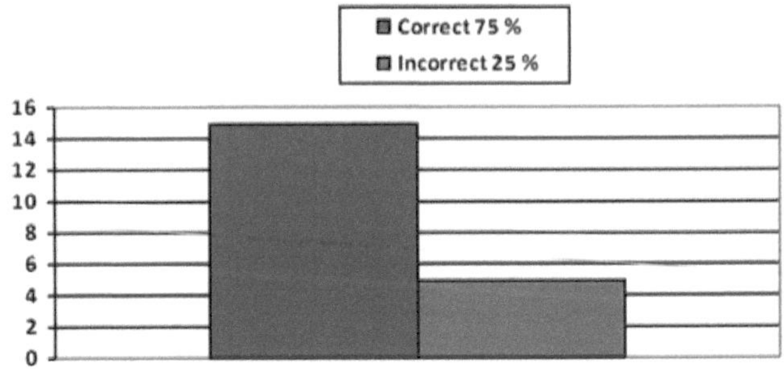


Again, a clear majority of correct indications and interpretations may be due to the fundamental importance of the up-down orientation constituting the primary metaphor underlying the sign.

5. Which of the following PJM signs expresses the concept of 'time' and why?

*Figures 24 and 25: The* PJM *signs for 'time' and 'ill' (ESLC, 2014).*

In sub-group A of the 14–16-year-olds, 6 out of 64 (9.38 %) participants correctly pointed out and interpreted the sign mentioning the idea of flow, which is the source domain of the primary metaphor Time Is Motion (Lakoff & Johnson, 1999, p. 52) underlying the sign. 4 participants only pointed out the correct sign. Most of the 54 wrong responses interpreted the second sign as the common time-indicating gesture pointing to a wrist-watch, in spite of there being a significant difference between the configuration and location of the hands in the gesture and the sign. In sub-group B, 2 out 9 (22.22 %) participants correctly pointed out and interpreted the sign. A detailed description of the hand-shape meaning was given by a participant that reported long-term contact with signed communication.

*Figures 26 and 27: 'Time' – sub-groups A and B of the 14–16-year-olds.*

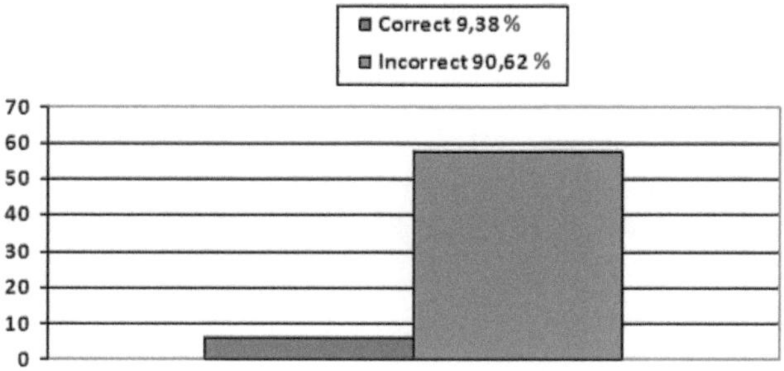


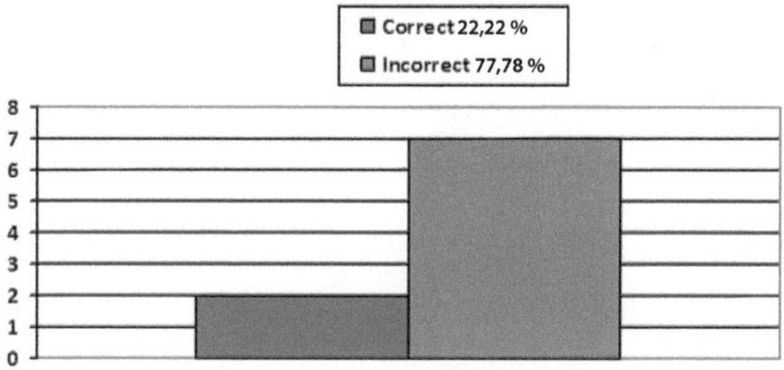


For the fourth time, the knowledge of common gesture influenced the selection and the interpretation of the sign. Its impact on the participants' responses was stronger than in any of the previous tasks.

10 out of 20 (50 %) of the 18–19 year-olds correctly pointed out and interpreted the concept of time's flow. 3 pupils additionally said that the c-hand shape represented the initial letter of the Polish word *czas* [Eng. *time*]. 1 participant did not do the task. Among the 9 wrong indications and interpretations, 4 mentioned the movement of clock hands; 1 pointed out the letter T as standing for *time*; 1 mentioned the loop of time; 1 provided no motivation at all.

*Figure 28: 'Time' – the 18–19-year-olds.*

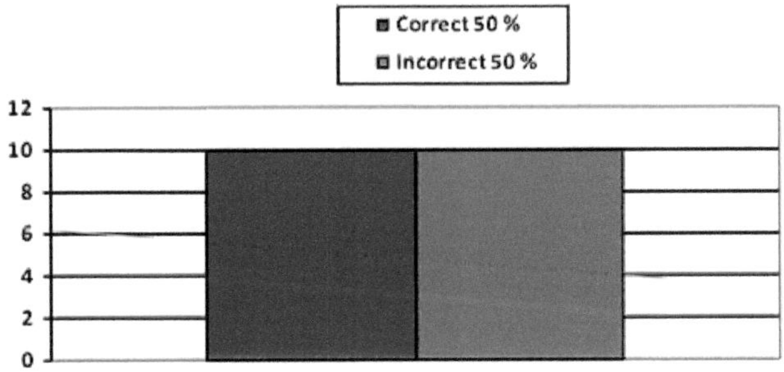


The result is close to the ones achieved for the previous signs in this age group. Except for the *T* hand-shape standing for *time*, little influence of gestures on the participants' responses was observed.

7. Which of the following PJM signs expresses the concept of 'promotion' and why?

*Figures 29 and 30: The* PJM *signs for 'promotion' and 'pregnancy' (ESLC, 2014).*

In sub-group A of the 14–16-year-olds, 29 out of 64 (45.31 %) participants correctly pointed out and interpreted the metaphorical movement up, which is the source domain of the primary metaphors More Is Up and Control is Up (Lakoff & Johnson, 1999, pp. 51, 55). 18 participants only correctly pointed out the sign. 17 pupils did not respond to the task at all. In sub-group B, 5 out of 9 (55,56 %) participants correctly pointed out and interpreted the sign. 1 participant correctly pointed out the sign. In 3 cases, there was no response to the task.

*Figures 31 and 32: 'Promotion'– sub-groups A and B of the 14–16-year-olds.*

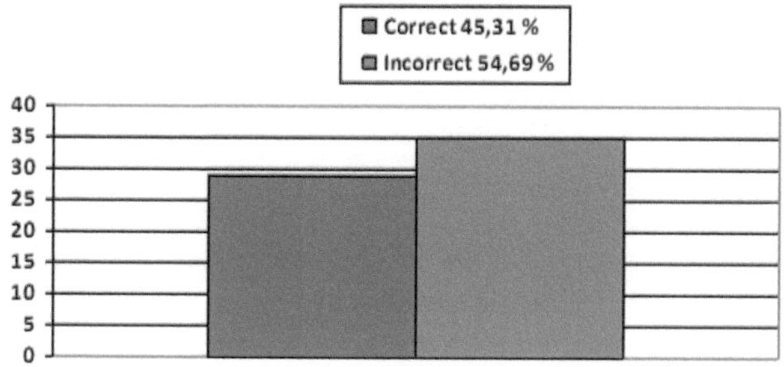


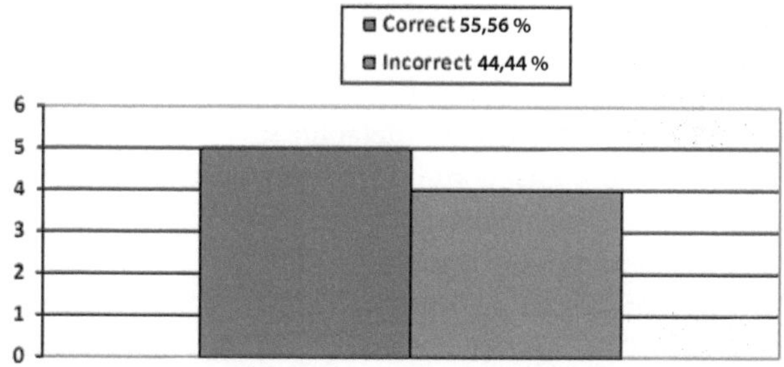


In both sub-groups, the number of correct indications and interpretations increased with the age of the participants, e.g., it was higher for the 16-year-olds than for the 14-year-olds.

19 out of 20 (95 %) of the 18–19-year-olds correctly pointed out and interpreted the sign.1 participant gave a correct interpretation, but pointed to no sign.

*Figure 33: 'Promotion' – the 18–19-year-olds.*

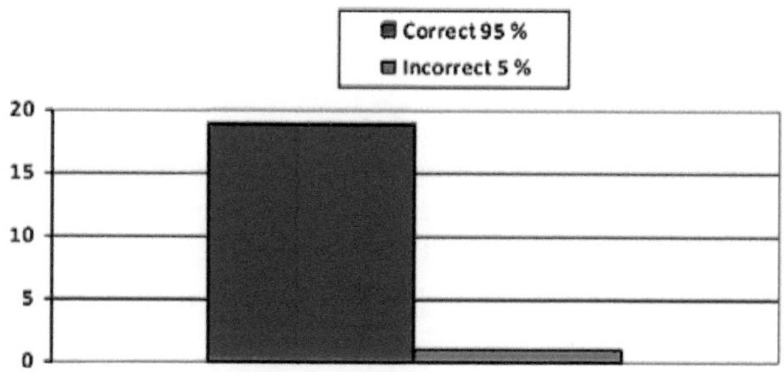

As in the younger age group, the absolute majority of indications and interpretations were correct. In both cases, the result may again be due to the fundamental importance of the up-down orientation being the source domain of the primary metaphors underlying the sign.

## 5. Conclusions

The results estimated by means of the number and percentage rate method for all tested groups lead to the following conclusions:

a/ the hearing have some comprehension of non-literal aspects of signed communication;

b/ the level of comprehension depends on age – as a rule, it is higher in older groups;

c/ non-literal aspects most straightforwardly perceived are related to the spatial orientations underlying some primary metaphors;

d/ in the case of four signs, common gestures present in diverse cultures strongly influenced the interpretations; in one case (Task 6), a big gap between the younger and the older groups was observed in the influence of the common gesture on the interpretation of the signs;

e/ the complexity of the underlying metaphors also influenced the results – relatively small number of correct interpretations were given for the complex metaphor The Mind Is a Machine;

f/ some interpretations mentioned other metaphors underlying the signs, e.g. the complex metaphor Thinking Is Moving in the *PJM* sign 'to think';

g/ some interpretations of the signs were based on metonymy: Typical Action for the Person in the *PJM* sign for 'court of justice', The Clock for Time and Part of a Form for the Whole Form (Radden & Kövecses, 1999, pp. 28, 36) in the *PJM* sign for 'time', as well as The Head for Thinking in the *PJM* sign for 'thinking'.

## References

*Białas, M.* (2007). *Głusi, język, metafora: Rozumienie metaforycznego znaczenia wyrażeń językowych przez uczniów niesłyszących* [The deaf, language, and metaphor: Comprehension of metaphorical linguistic expressions by deaf children]. Piotrków Trybunalski, Poland: Naukowe Wydawnictwo Piotrkowskie.

*Brennan, M.* (1990). *Word formation in British Sign Language*. Stockholm: University of Stockholm.

*Czajkowska-Kisil, M.* (2008). Pułapki metodologiczne w badaniach *PJM* [Methodological problems of research on Polish Sign Language]. In E. Twardowska

(Ed.), *Stan badań nad Polskim Językiem Migowym* [Research on Polish Sign Language – state of the art] (pp. 11–15). Łódź, Poland: Polski Związek Głuchych Oddział Łódzki.

ELSC/European Sign Language Centre. (2014). *Sign language dictionary.* Retrieved June 3, 2014 from the European Sign Language Centre at http://www.spreadthesign.com/pl.

*Grady, J. E.* (1997). *Foundations of meaning: Primary metaphors and primary scenes.* PhD dissertation, Linguistics Department, University of California, Berkeley, CA.

*Grote, K., & Linz, E.* (2003). The influence of sign language iconicity on semantic conceptualization. In W. G. Müller & O. Fischer (Eds.), *From sign to signing: Iconicity in language and literature 3* (pp. 23–40). Amsterdam: Benjamins.

*Grzymska, A.* (1993). Kognitywne ujęcie metafor w PJM [A cognitive view of metaphor in Polish Sign Language]. In E. Twardowska (Ed.), *Stan badań nad Polskim Językiem Migowym* [Research on Polish Sign Language – State of the Art] (pp. 87–95 ). Łódź, Poland: Polski Związek Głuchych Oddział Łódzki.

*Hendzel, J. K.* (1995). *Słownik polskiego języka miganego* [A dictionary of Polish signs]. Olsztyn: Offer.

*Lakoff, G., & Johnson, M.* (1980). *Metaphors we live by.* Chicago: University of Chicago Press.

*Lakoff, G., & Johnson, M.* (1999). *Philosophy in the flesh: The embodied mind and its challenge to western thought.* New York: Basic.

*Landauer, T. K., & Dumais, S. T.* (1997). A solution to Plato's problem: The latent semantic analysis theory of acquisition, induction, and representation of knowledge. *Psychological Review, 104,* 211–240.

*Lane, L. G.* (1990). *Gallaudet survival guide to signing.* Washington, DC: Gallaudet University Press.

*Pizzuto, E., Boyes-Braem, P., & Volterra, V.* (1996, September). *Seeing through signs' iconicity: A crosslingustic-crosscultural study of signers and speakers.* Fifth International Conference on Theoretical Issues in Sign Language Research, Montreal, Quebec, Canada.

*Podgórska-Jachnik, D.* (2004). *Przekaz pantomimiczny w komunikacji z dzieckiem niesłyszącym* [Pantomime-based communication with deaf children]. Łódź, Poland: Wydawnictwo Uniwersytetu Łódzkiego.

*Radden, G., & Kövecses, Z.* (1999). Towards a theory of metonymy. In K.-U. Panther & G. Radden (Eds.), *Metonymy in language and thought* (pp. 17–59). Amsterdam: Benjamins.

*Ritchie, L. D.* (2013). *Metaphor.* Cambridge: Cambridge University Press.

*Smith, C.* (2010). *Sign language companion: A handbook of British signs*. London: Souvenir.

*Sutton-Spence, R., & and Woll, B.* (2010). *The linguistics of British Sign Language: An introduction*. Cambridge: Cambridge University Press.

*Taub, S. F.* (2001). *Language in the body: Iconicity and metaphor in American Sign Language*. Cambridge: Cambridge University Press.

*Wilbur, R. B.* (1987). *American Sign Language: Linguistic and applied dimensions*. Boston, MA: College-Hill.

*Wilcox, P. P.* (2000). *Metaphor in American Sign Language*. Washington, DC: Gallaudet University Press.

*Wilcox, S.* (2008). Signed languages. In D. Geeraerts & H. Cuyckens (Eds.), *The Oxford handbook of cognitive linguistics* (pp. 1113–1136). Oxford: Oxford University Press.

*Wilcox, S., Wilcox, P. P., & Jarque, M. J.* (2003). Mappings in conceptual space: Metonymy, metaphor, and iconicity in two signed languages. *Jezikoslovlje*, *4* (1), 139–156.

Agnieszka Stępkowska
*University of Social Sciences, Warsaw, Poland*

# Second language acquisition in the canton of Zurich: the Swiss are fond of English

**Abstract:** The paper addresses the multilingual policy of the Swiss in the canton of Zurich with reference to the partner-language model offered by their educational system. The outlined case study of Zurich builds on the theoretical assumptions of contact linguistics and macro sociolinguistics, notably the sociology of language introduced by Joshua Fishman (1972, 1997). The paper will present a relevant part of the collected material together with conclusions that formed a bigger CATI survey carried out in the canton of Zurich in 2011. The collected interviews, based on a questionnaire, have made it possible to outline the language repertoires of the Swiss, as well as their opinions and attitudes towards English and its acquisition. The results from the survey reveal language behaviour in a multilingual context marked by the presence of a global language.

Keywords: multilingualism, Switzerland, language policy, telephone interviews, language attitudes.

## 1. Multilingual policy: an outline

Language policy is a broad term that refers to the decisions about language rights and access to languages, including the roles and functions of languages and their varieties in individual countries (Phillipson and Skutnabb-Kangas, 1996). The Swiss language policy, in particular, seeks to solve communication problems by providing support for linguistic pluralism (Blount and Sanches, 1977; Daoust, 1997). The system of education is decentralized and, thus, differs from canton to canton. All cantons share the federative obligation of a second official language teaching (Schmid, 2001). The idea is that every Swiss should have an active command of a second official language, and at least passive knowledge of a third official language of his/her country. A first foreign language is obligatory, with the reservation that the choice of this language cannot be the one between a 'Swiss' official language and English. The most obvious choice of a second national language (and simultaneously a first foreign language) for the three language groups of the Swiss – French, Italian and Romansh – is German. In turn, the largest linguistic group of the German-speaking Swiss learn French as their second language. Most often for the Italian-speaking Swiss, German is a first foreign language and French a second one (see also Pap, 1990; Watts, 1997). The choice of the third official language, or a second foreign language, is usually Italian, although recently it has been almost entirely supplanted by English, which is more and more appreciated by the Swiss (Dürmüller, 1986, 1991, 1992, 1994).

The main part of language policy is language planning. A closer look at the language planning in Switzerland is intriguing due to the language complexity, i.e. three official, or four national languages. This language diversity has been intensified by the fact that the majority of the population cultivates a range of mutually intelligible and mainly spoken Alemannic dialects, including five idioms within the Romansh language which, in turn, often lack the required intelligibility needed for effective communication. Everyday life in the German-speaking area of Switzerland is dominated by Alemannic dialects, while the written form and certain official contexts have been reserved for the standard German (Porębski, 1994; Szulc, 1999). This variety of functions fulfilled by respective language varieties is typical of diglossia, a phenomenon originally described by Ferguson (1959). For Phillipson and Skutnabb-Kangas (1996), language planning is essential in the multilingual and multidialectal milieu. Above all, it is a demonstration of the political and economic choices of those who are held responsible for language planning. Hence, language planning as a rule aims to obtain extra-linguistic goals, such as consumer protection, scientific exchange, national integration, political control, economic growth, minority pacification, as well as mass mobilization of the national or political movements.

## 2. Partner-language model and English

The Swiss model of intra-national communication stipulates that the Swiss speak their mother tongues and are understood by fellow countrymen speaking other languages. Besides the mother tongue, a second national language is taught to reach receptive and productive levels necessary for active communication, whereas a third national language should be mastered up to a passive level, sufficient for understanding (Widmer et al., 1987, p. 101). However, these efforts are not tantamount to common bi- or trilingualism. Most Swiss know only one language. Probably fewer than half can speak or write in a second national language, revealing diverse levels of fluency. Instead, a growing number of them choose to learn English (Pap, 1990; Watts, 1991). The intra-national communication is expected to work by the so-called 'partner-language model' (Dürmüller, 1992). Promoted by the Swiss educational policy, bilingualism would be a satisfactory solution merely for the two largest language groups. In the case of language contact, someone needs to have both active and passive knowledge of the other language. One of the two languages needs to be chosen as the medium of communication. In the contact between the German- and French-speaking Swiss, either group may theoretically use their mother tongue (L1). It may be assumed that the level of their passive competence in the other language (L2) will make a conversation possible. However, this assumption is far-fetched, because most German-speaking Swiss do not use the German that is taught to the French-speaking Swiss at schools. In fact, the mother tongue of the Swiss from the German-speaking part of Switzerland are Alemannic dialects, and they themselves learn the standard variety of German at school.

The intra-national communication in quadrilingual Switzerland, according to the above-mentioned partner-language model, is feasible in a situation when every potential interlocutor has in their individual repertoires three national languages. Dürmüller (1989, p. 5) gives a detailed account of ideal language combinations in the repertoires of the Swiss from the three official language groups, and he sketches out the potential 'threat' to those combinations in the form of English taking up the L2 position in each language group. The Swiss educational system, guided by the assumptions of the partner-language model, aims to equip every citizen with at least rudimentary knowledge of a second official language. However, this strategy bears discrepant results. Many graduates of the secondary or tertiary educational level demonstrate, at best, only a passive command of a second official language.

English has become one of the basic skills needed to acquire the more advanced and specialized knowledge necessary in the future. In other words, English is now necessary to "learning how to learn" (Graddol, 2006, p. 72). Besides, learning English, with its roles and functions, adds to the complexity of language behaviours in multilingual societies. It would seem logical and purposeful to introduce a law, especially in an officially multilingual country like Switzerland where three national languages are big European languages, stipulating that a first 'foreign' language taught at schools must be one of the other two national languages. Yet, since none of the three languages, i.e. German, French and Italian, is 'foreign' in Switzerland, they are referred to by a French expression as *langue nationale 2* (national language 2). As to English in Switzerland, there are three criteria specifying the status of a foreign language that are worth recalling after Prcic (2003, p. 35): (1) it is not the native/ first language of a country, (2) it is not an official language of a country, and (3) it is taught in schools as a subject.

## 3. Methodological note: research technique, sample and aim

The research was conducted from 28 June to 20 July 2011, from the CATI telephone studio centre of the PBS Ltd research institute in Sopot, Poland. The CATI technique (i.e. Computer Assisted Telephone Interviewing) was selected as the method of data collection. CATI is a technique used to realize large quantitative research projects. It consists in conducting interviews over the telephone aided by the use of the computer. In total, 400 successful phone interviews were made, based on a questionnaire with closed-ended questions. The screening index, i.e. the number of connections needed for one effectively conducted interview, equalled 14.1 connections. The mean time of one successful interview was 6.6 minutes. For the research sample of a size of n=400, the maximum error of estimation amounted to nearly 5% at the assumed confidence level of 95%. The research was based on probability sampling, representative of the city and the canton of Zurich, and characterized on the basis of the data concerning gender, age, education and employment. The number of women and men participating in the research is comparable. All respondents were adults, half of whom are persons between 35 and 54 years old. The group of respondents over 55 equals 38%. The youngest age category, i.e. persons between 18 and 34, made up as many as 11%. Nearly half of the respondents are people with primary or lower secondary education (48%). The second biggest group concerns the graduates of universities or colleges (28%). Every fifth respondent has declared to have an upper secondary level of education (20%). And, two thirds of all interviewees work professionally (67%).

The aim of the CATI survey is to focus on data such as the knowledge of English and other foreign languages, the years of English instruction received, the preferred language to speak, the importance of learning English, and the interest and motivation to improve English language skills. The respondents were also asked to assess the future of English in the canton of Zurich, and to express their opinions as to whether the knowledge of English may be regarded as an asset in Switzerland. In other words, the aim of the survey – for the purpose of this article – is to analyze the language behaviour of the German-speaking Swiss by outlining their attitudes and opinions about the acquisition of English in the context of their four national languages.

## 4. Research results[1]

Most inhabitants of the canton of Zurich (78%) point to the German language as their mother tongue. From among the group whose mother tongue is some other language than English, nearly 90% of the respondents had learnt it at some time.

---

1 Based upon research results published in Stępkowska (2013: 240–246). Here are presented only fragmentary results of the CATI research conducted in the canton of Zurich, i.e. those pertaining to the concept of this article. The entire research, the questionnaire on which it was based, as well as the analysis and results placed in a broader multilingual context, have been published in my latest book (2013) *Multilingualism*

English is learnt somewhat more often by women than by men (90% vs. 85%). Learning English is correlated with the age of the respondents – as many as 98% from the youngest group aged between 18–34 have declared that they had English instruction. An analogical tendency concerns the level of education – the higher it is, the more persons have learnt this language. Interestingly, the professional situation of the respondents (working / non-working) does not reveal differences with regard to this problem. The biggest group of the inhabitants of the canton of Zurich admitted to having studied English for between 3 to 6 years.

Almost 80% of all respondents can speak French, and half of them know Italian. As regards these two languages, women are somewhat ahead of men in number (83% vs. 75%). Respondents aged between 18–34 most often know French (88%), whereas Italian has been indicated only by every fourth person from this age group (26%). When asked about the knowledge of other foreign languages, the inhabitants of the canton of Zurich in the first instance pointed to Spanish, Russian and Portuguese.

The everyday communication in German is the preference of 70% of all residents in the canton of Zurich. However, noteworthy is the fact that every eighth respondent (12%), when asked about the language they like to speak most, has indicated English. French and Italian have scored but trace quantities of answers (4% and 3% respectively). Women seem to resort to English more willingly than men (16% vs. 7% respectively). The respondents from the oldest age group (aged 55 or more) pointed to English as their most preferable language of communication less frequently than the other age groups (only 8%).

The need to learn English is essential according to 89% of the inhabitants of the canton. Half of all the respondents would like to know English better, and particularly those from the youngest age group (70%). The significant role of English is backed up by the respondents' opinions that the knowledge of English is essential (52%) or quite important (44%). According to two thirds of the respondents, the future of English will be better than today, i.e. the role of this language will be increasing. One fourth of the interviewed persons believe that the situation of English in the canton of Zurich will not be much different from the present one, and less than a mere 2% think that it will be worse than today.

Apart from answering the questions, the respondents were asked to take a stance on seven statements that were included in the final part of the questionnaire. The respondents were asked to voice their opinions concerning the role of

---

*and English: The canton of Zurich as a linguistic paradigm.* Frankfurt am Main: Peter Lang Verlag.

English. Almost all the respondents (97% of overall answers 'definitely yes' and 'somewhat yes') agreed that they would, actually or hypothetically, provide their children with English instruction. A similar percentage (95%) of the respondents stated that a knowledge of English is regarded as an asset in Switzerland. Almost 90% of interviewees declared that if they did not learn English at school or university, they would try to learn it on their own. The same percentage of respondents are convinced that English is a link that connects them with the outside world, and that English reinforces interpersonal contacts (89%). A smaller proportion of the respondents, yet still big enough to be regarded as noteworthy (as many as 52%!), agreed with the statement that English is more useful in Switzerland than any of the Swiss national languages. The least number of respondents identified themselves with the statement that English is a link that connects them with their fellow countrymen from different language areas of Switzerland (38%).

It has turned out that as many as 86% of the persons interviewed are of the opinion that the language issues mentioned in this research are 'very important' or 'quite important.' Thus, it may be concluded that for almost 90% of the respondents aged between 35 and 55, i.e. persons who are most work-efficient, language issues are of considerable significance in their lives. This may be indicative of the fact that an appreciable percentage of the Swiss living in the canton of Zurich are aware of the language processes occurring in their society, and that the question of languages in a multilingual country does not seem to lose timeliness and forms an important constituent of public life. Curiously enough, the importance of language issues is felt equally by the working and non-working respondents.

The research results indicate that English enjoys popularity and prestige among the German-speaking inhabitants of the canton of Zurich. Only every tenth person when asked about their experience with English lessons stated that they had nothing to do with this language. Practically the entire young generation of the Swiss can hardly imagine their lives without the knowledge of English. Language pragmatism in the respondents' attitudes was particularly detectable in their assessment of the role of English in their lives. As many as nine out of ten persons expressed their concern not only with their own language education, but also with the English tuition of their children. In turn, the conviction that English has the biggest communication potential worldwide is undoubtedly a solid motivational base for 90 per cent of the respondents. Also, most respondents believe that English will gain significance in the near future. Nevertheless, to state that English functions as a *lingua franca* within the country would be an overstatement, as the Swiss from different language areas communicate still more often in one of their official languages rather than in English. Yet, regardless of the chosen language, the communication that fulfils the needs of interlocutors always proves their language pragmatism.

## 5. Concluding remarks

This article has aimed to outline the position of English in the multilingual policy of Switzerland. Language planning concerning language acquisition is a political act directed to solve language problems in society since the social effects of language problems and the social context, to some extent, lead to disciplined decisions about language on the national level (Jernudd and das Gupta, 1971; also Lotherington, 2004). Following Bourdieu (1986), language may be subjected to planning as it is a resource that undergoes evaluation. The effects of language planning are more predictable if the social context is taken into account, and especially the cultural and economic forces which contribute either to language preservation or language shift. The efforts of language planning may be successful if supported by economic or social benefits addressed to minorities (Linder, 1996; Paulston, 1988).

It has been observed that the Swiss barely manage to acquire a sufficient level of bilingualism nationwide (Dürmüller, 1991, 1992). It might seem that the partial reason for this situation is the impact of English in Switzerland because the efforts invested in second language acquisition are channelled in two directions. On the one hand, young people learn a second national language at school as it is mandatory there; and on the other, English is learnt voluntarily and with strong motivation. English is considered fashionable and more useful than any of the 'Swiss' national languages. The increasing use of English in Switzerland counteracts the efforts of language planning that aim to reinforce the multilingual dialogue among the Swiss. It appears that the decisive factor in language acquisition is not the political will of relevant institutions and authorities, but the specific needs of individuals. The case of Switzerland proves that in a multilingual society the status of a language of wider communication may be most easily acquired by a foreign language. Although English is merely a foreign language in Switzerland, it gains the upper hand over the national languages of this country due to its neutrality – which helps avert conflicts among the diverse language groups. English is readily accepted in situations where the partner-language model fails. Today Switzerland seems to oscillate between two systems, i.e. the traditional partner-language model that provides for satisfactory L2 competence, and the choice of the language of wider communication (English) in the fields of life which require the use of a specialized language. Thus, Switzerland needs to prepare for its multilingual future, but with a slightly changed linguistic makeup, i.e. with national languages coexisting with English.

There is also an ongoing debate that concerns the language choice in schools in the educational systems of the Swiss cantons. Such discussions are emotional as they include not only the status of languages in the public life of Switzerland but, more importantly, language teaching in the public educational system (Watts, 2001). The point is that language acquisition should be additive and not subtractive. In the learning process the repertoires of individual persons should widen. New languages, including lingua francas, should not be taught at the cost of national languages (see Skutnabb-Kangas, 2003). The presence of English in the Swiss public space undergoes a gradual transformation. In the past the language repertoires of individuals were formed almost exclusively within the three national languages of Switzerland, both in the case of bilinguals or, though more infrequently, trilinguals. English was absent from these repertoires. At present English has no official recognition in Switzerland, although it is already there. There are no regulations which would demand a limitation or a ban on English at the Swiss universities. Besides, universities do not have any clear policy in this respect, which may result in a supposition that English is now making big strides within the academic circles in Switzerland (Dürmüller, 2001; more in Murray and Dingwall, 1999, p. 192).

## References

*Blount, B. G., & Sanches, M. (Eds.)* (1977). *Sociocultural dimensions of language change.* New York – San Francisco: Academic Press.

*Bourdieu, P.* (1986). The forms of capital. In J. G. Richardson (Ed.), *Handbook of theory and research for the sociology of education* (pp. 241–258). New York: Greenwood Press.

*Bourdieu, P.* (1991). *Language and symbolic power.* Cambridge: Polity Press.

*Daoust, D.* (1997). Language planning and language reform. In F. Coulmas (Ed.), *The handbook of sociolinguistics* (pp. 436–452). Oxford: Blackwell.

*Dürmüller, U.* (1986). The status of English in multilingual Switzerland. *Bulletin CILA, 44,* 7–38.

*Dürmüller, U.* (1989). Attitudes towards English as a possible lingua franca in Switzerland. *York Papers in Linguistics, 14,* 3–17.

*Dürmüller, U.* (1991). Swiss multilingualism and intra-national communication. *Sociolinguistica, 5,* 111–159.

*Dürmüller, U.* (1992). The changing status of English in Switzerland. In U. Ammon & M. Hellinger (Eds.), *Status change of languages* (pp. 355–370). Berlin – New York: Mouton de Gruyter.

*Dürmüller, U.* (1994) Multilingual talk or English only? The Swiss experience. *Sociolinguistica, 8,* 44–64.

*Dürmüller, U.* (2001). The presence of English at Swiss universities. In U. Ammon (Ed.), *The dominance of English as a language of science: Effect on other languages and language communities* (pp. 389–403). Berlin – New York: Mouton de Gruyter.

*Dürmüller, U.* (2002). English in Switzerland: From foreign language to lingua franca. In D. Allerton, P. Skandera & C. Tschichold (Eds.), *Perspectives on English as a world language* (pp. 115–123). Basel: Schwabe.

*Ferguson, Ch.* (1972). Diglossia. In P. P. Gigliolo (Ed.), *Language and social context. Selected readings* (pp. 232–251). Harmondsworth: Penguin Books. (Original work published 1959)

*Fishman, J. A.* (1972). *The sociology of language: An interdisciplinary social science approach to language in society.* Rowley: Newbury.

*Fishman, J. A.* (1997) The sociology of language. In N. Coupland & A. Jaworski (Eds.), *Sociolinguistics: A reader and coursebook* (pp. 25–30). London: Macmillan Press.

*Graddol, D.* (2006). *English next: Why global English may mean the end of 'English as a foreign language'.* London: British Council.

*Jernudd, B.,* & *Das Gupta, J.* (1971). Towards a theory of language planning. In J. Rubin & B. Jernudd (Eds.), *Can language be planned?* (pp. 195–215). Honolulu: University of Hawaii Press.

*Linder, W.* (1996). *Demokracja szwajcarska: Rozwiązywanie konfliktów w społeczeństwie wielokulturowym.* Rzeszów: Wyższa Szkoła Pedagogiczna.

*Lotherington, H.* (2004). Bilingual education. In A. Davies & C. Elder (Eds.), *The handbook of applied linguistics* (pp. 695–718). Oxford: Blackwell.

*Murray, H., & Dingwall, S.* (1999). The future of English in Switzerland: A majority/minority problem?. *Bulletin Suisse de Linguistique Appliquée, 69* (2), 189–206.

*Pap, L.* (1990). The language situation in Switzerland: An updated survey. *Lingua, 80,* 109–148.

*Paulston, Ch. B. (Ed.)* (1988). *International handbook of bilingualism and bilingual education.* New York: Greenwood Press.

*Phillipson, R., & Skutnabb-Kangas, T.* (1996). English only worldwide or language ecology?. *TESOL Quarterly, 30* (3), 429–452.

*Porębski, Cz.* (1994). *Na przykład Szwajcarzy... Eseje i rozmowy.* Kraków: Znak.

*Prcic, T.* (2003). Is English still a foreign language?. *The European English Messenger, 12* (2), 35–37.

*Schmid, C. L.* (2001). *The politics of language: Conflict, identity and cultural pluralism in comparative perspective.* Oxford: Oxford University Press.

*Skutnabb-Kangas, T.* (2003). Linguistic diversity and biodiversity. The threat from killer languages. In Ch. Mair (Ed.), *The politics of English as a world language: New horizons in postcolonial cultural studies* (pp. 31–52). Amsterdam - New York: Rodopi.

*Szulc, A.* (1999). *Odmiany narodowe języka niemieckiego: Geneza – rozwój – perspektywy.* (Rozprawy wydziału filologicznego, Tom LXX.) Kraków: PAU.

*Watts, R.* (1991). Linguistic minorities and language conflict in Europe: Learning from the Swiss experience. In F. Coulmas (Ed.), *A language policy for the European community: Prospects and quandaries* (pp. 75–101). Berlin: Walter de Gruyter.

*Watts, R.* (1997). Language policies and education in Switzerland. In R. Watts & J. Smolicz (Eds.), *Cultural democracy and ethnic pluralism: Multicultural and multilingual policies in education* (pp. 271–302). Frankfurt am Main: Peter Lang.

*Watts, R.* (2001). Discourse theory and language planning: A critical reading of language planning reports in Switzerland. In N. Coupland, Ch. Candlin & S. Sarangi (Eds.), *Sociolinguistics and social theory* (pp. 297–320). London: Addison Wesley Longman.

*Widmer, A., Laubscher, C., & Fluegel, Ch.* (1987). *Herausforderung Schweiz. Materialien zur Förderung des Unterrichts in den Landessprachen.* Bern: Schweizerische Konferenz der kantonalen Erziehungsdirektoren.

# Part Two: Language Corpora

Valérie Bourdier
*University of Paris-Est (UPEC – EA 3958), France*

Agnès Leroux
*University of Paris-Ouest Nanterre, France*

# *No problem* or *no problems*? Special problems raised by the reference to absence in the sequences *no+N-Ø* and *no+N-s*

**Abstract:** This paper enquires into the field of noun determination and focuses on the construction '*No* + Noun', as occurs in the following excerpt:

> *(…) and the executive government of the United States, […], will recognize and maintain the freedom of such persons and will do no act or acts to repress such persons, or any of them, in any efforts they may make for their actual freedom.* (Lincoln's Proclamation Act, 1862)

The existence of the two sequences, *no act* and *no acts*, raises the question of the specificity of noun determination in negative structures in English. Insofar as the two structures are used in the same sentence, it may be hypothesized that they convey differences in meaning. The aim thus is to examine examples in which the two constructions are grammatically allowed in this pattern, and to elucidate the process through which reference and meaning are construed. We will demonstrate through corpus-based study, led within the frame of the Theory of Enunciative Operations, that, although they seem to be interchangeable in some contexts, they are not synonymous.

We will address these issues via the examination of examples taken from corpora of contemporary written English, such as the COCA.

Keywords: Linguistic operation, negation, reference to absence, noun determination, tense, aspect, generic/specific, quality/quantity.

## Introduction

This paper shall address the issue of the sequence *no + countable noun* insofar as the countable noun following *no* can occur either in the singular or in the plural.

Most authors (such as M. Swan 1995: 371) contend that after *no*, countable nouns are usually plural unless the sense makes a singular noun necessary. M. Swan gives the following examples:

(a) *He's got no children.* (rather than *He's got no child.*)
but
(b) *He's got no wife.* (rather than *He's got no wives.*)
meaning that he would be expected to have only one wife if he had.

However, sequences such as the one in the following excerpt are often encountered:

(1) *MADIGAN: I don't work for the school district, but I do know this,* ***no teacher*** *has complained to the teacher's union, and the video was shown over a month ago. I find it hard to believe that the board had any real issue with it. HILL: Board members are expected to take up this issue of Mitchell's future, Tuesday night. (END-VIDEOTAPE) S-O'BRIEN: That story from Reporter Darlene Hill, from CNN affiliate, WFLD, reporting. Interesting there, huh? M-O'BRIEN: Yes, I can see how the humor might be misconstrued. S-O'BRIEN: But* ***no teachers*** *complained.*
(COCA: CNN-AM, 2006)

The co-occurrence of *no teacher* and *no teachers* in the same excerpt raises the question of the specificity of noun determination in negative structures. Insofar as the two structures are used in the same context, we may hypothesize that they convey differences in meaning and that they may not be interchangeable.

This analysis will be carried out within the theoretical framework of the Theory of Enunciative Operations as set out by A. Culioli. Proceeding from the methodological principle of using a theory of observables, i.e. a data-driven approach, it gathers locally comparable observations in order to re-construct the formal mechanisms that underlie the production and recognition of interpretable '*shapes*', namely utterances (Culioli 1990:178). After a quantitative approach, which will lead us to identify the recurrent structures at stake, we will focus on an in-depth analysis of contextual parameters.

## 1. A quantitative approach to the problem

We will start with the outcome of some research within the COCA corpus, which shows that there is a clear significant overall tendency towards more frequency of the combination *no + singular noun*:

The first thirty-two nouns are singular (even if they are not all countable nouns), and the eight most frequent combinations are: *no way*, *no idea*, *no reason*, *no question*, *no doubt*, *no time*, *no problem*, *no need*; the first noun found in the plural is *sign* (*no signs*) and ranks 33rd.

The following table testifies to those numbers:

*Table 1: Opposition between the first eight plural nouns and their corresponding singular forms*

| No + N-Ø | results | rank | No + N-*s* | results | rank |
|---|---|---|---|---|---|
| ***No sign*** | 2922 | **14** | *No signs* | 1219 | **33** |
| **No plan** | *415* | ***121*** | No plans | *1187* | ***35*** |
| ***No child*** | 1416 | **26** | *No children* | 898 | **45** |
| ***No problem*** | 5136 | 7 | *No problems* | 835 | **52** |
| ***No word*** | 786 | **57** | *No words* | 723 | **64** |
| ***No difference*** | 1836 | **18** | *No differences* | 650 | **73** |
| ***No question*** | 6669 | **4** | *No questions* | 633 | **76** |
| **No regret** | *48* | ***968*** | No regrets | *593* | ***80*** |

However, in order to make our study more consistent, we have voluntarily focused our attention on eight English nouns displaying various frequency rates, three of which being more frequent in the plural combination: *dog*, *eye*, *cloud*.

*Table 2: Comparison of frequency rates*

| | No + N-Ø | | No + N-s | |
|---|---|---|---|---|
| **No man / No men** | 1455 | (24th) | 122 | (419th) |
| **No child / No children** | 1416 | (26th) | 898 | (45th) |
| **No fear / No fears** | 767 | (60th) | 25 | (1655th) |
| **No woman / No women** | 448 | (112nd) | 336 | (153rd) |
| **No teacher / No teachers** | 87 | (572nd) | 35 | (1274th) |
| ***No dog / No dogs*** | *91* | *(552nd)* | *115* | *(449th)* |
| ***No eye / No eyes*** | *80* | *(639th)* | *103* | *(498th)* |
| ***No cloud / No clouds*** | *17* | *(2185th)* | *65* | *(742nd)* |

This comparative grid will lead us to our qualitative study of some items in the data to elaborate on the different meanings constructed by the patterns under scrutiny.

## 2. A qualitative study carried out within the enunciative approach

The noun *cloud*, which ranks 2185th in the singular and 742nd in the plural form, is used in the plural in example (2) and in the singular in example (3):

> (2) *That afternoon of July 29, 1878, was exactly like a thousand others that Dirk Skye had experienced, except that the light was dying. He glanced out the schoolhouse window and saw nothing amiss. There were* ***no clouds.*** *He returned to his lessons. This day he had seven students in his Wind River Reservation school. [...]. They sat at their desks staring at him in the deepening dusk.*
> (COCA: Richard S. Wheeler, *The owl hunt*, New York: Forge Edition, 2010)

> (3) *It was a late summer afternoon, warm, still, and he swore* ***no cloud*** *crossed the sky. As he hammered a row of asphalt shingles, a shaft of lightning struck a chimney at the end of the three-story building and knocked loose a level of bricks. Electric shock burned through his fingers and soles of his shoes.*
> (COCA: William Hoffman, *Prodigal*, Sewanee Review Vol.104 Issue 1, 1996)

We shall proceed to a comparative linguistic analysis of the way the reference to absence is constructed in each instance.

In (2), the noun phrase *no clouds* is included in a sequence which refers to a fundamental linguistic operation linked to predicating existence, as marked by the existential expression *there were*. This reference to the spatio-temporal location of clouds, i.e. their situational existence in space and time (or absence here) constitutes the Quantitative delimitation.

The determiner *no* negates the class of elements *clouds* in its totality, and the combination with *there were* results in the predication of the complete absence of occurrences of clouds. Absence is clearly opposed to presence since the presence of a number of occurrences of clouds was expected, with *the light was dying* paving the way for *there were clouds*.

In example (3), the negative noun phrase occurs as a syntactic subject with the noun *cloud* in the singular. At first sight, the two structures (*no clouds* and *no cloud*) may sound perfectly synonymous. However particular attention should be drawn to the context. For in (3) the possible presence of clouds has not been envisaged by the subjective origin of the discourse. In fact, what is negated is not so much the presence of clouds as such, as the very properties covering the notion[1] /CLOUDNESS/, in other words any element epitomizing the property 'be-cloud'.

1 "A notion is a complex system of representation organizing physico-cultural properties of a cognitive nature. The notion exists before words enter into categories; it is a generator of lexical units. It is generally represented between slashes, for example /CAT/;

The parallel existence of the two constructions should thus induce one to pay particular attention to the subjective and semantic features working as contextual parameters. These can yield insight into the process of the construction of meaning and might account for the singular or plural form after *no*, as illustrated in examples (4) and (5):

> (4) *For some reason, I expected something deep and formal, like Sitting Bull in council. I explain that I'm in the process of facing my biggest fears, and that I'd like to hire him to help me confront the greatest one of all: the fear of death. He laughs. "I have **no fears.** I've been shot out of cannons, been in knife fights.... Fear isn't in my blood. I'm Indian. All these silly people with their phobias.... Sure, I'll do it. For $1,500."*
> (COCA: Joe Kita, *This Man is Ready to Die*, Mens Health Vol.17 Issue 9, 2002)

> (5) – *You made me confess **the fears** that I have. But I will tell you also what I do not fear. I do not fear to be alone or to be spurned for another or to leave whatever I have to leave. And I am not afraid to make a mistake, even a great mistake, a lifelong mistake, and perhaps as long as eternity too.*
> *Cranly, now grave again, slowed his pace and said:*
> *- Alone, quite alone. You have **no fear** of that. And you know what that word means? Not only to be separate from all others but to have not even one friend.*
> (COCA: J. Joyce, *A Portrait of the Artist as a Young Man*, 1916)

In (4) the subject first takes into account all the elements of the class '*fears*' (*my biggest fears, the greatest one of all*) and rules them out with '*I have no fears*'. The class 'fears' has been constructed in the context and the character (here, the subjective origin of the discourse) can state that none of the different elements of the class corresponds to any feeling he may experience. Then, and only then, can he state that he has nothing to do with fear in its nature, hence the use of *fear* in the singular further down, with *Fear isn't in my blood*, which amounts to resorting to the notion itself.

In (5) the character first distinguishes between various occurrences of the notion /FEAR/, to finally assert the absence of them all (*be alone, be spurned for another, leave whatever I have to leave, make a mistake*). As a result, a shift is operated from the extraction of various occurrences of the notion /FEAR/ (various types

---

the notion is potentially a predicate ("being [a] cat", "having the property of being [a] cat"), defined exclusively in intension. Though possessing a certain degree of stability that is necessary for communication, notions are not fixed rigidly once and for all: on the contrary, they are dynamic and vary from one individual to another and from one situation to another. A notion can only be apprehended through its OCCURRENCES, whether they be phenomenal or enunciative."
(http://www-01.sil.org/linguistics/glossary_fe/defs/TOEEn.pdf)

of '*fear*') to an overall representation comprising their properties and that may be designated as '*fear*' in the singular. Hence ruling out all the occurrences amounts to discarding the notion itself. This procedure of abstraction seems to be necessary at this point, since in the context all the types of fear referred to prove irrelevant. As a result, the speaker cannot ascribe a set of properties to the representation that he has of real '*fear*' and deserving the name '*fear*'.

## 3. Differences in the reference to absence

This fundamental qualitative operation is particularly compatible with generic reference, as shown in examples (6), (7) and (8):

(6) *She was thinking that a boy should have the chance to do a few foolish things, get into a few wild scrapes, before he was asked to take on the responsibilities of a man. Hard on that thought came a rush of anger for the parents who had failed him, and a society which had carelessly thrust him beyond the pale. In some obscure way, she recognized that her feelings were inextricably bound up with the fierce, protective emotions she experienced for her son.* ***No child*** *should ever be alone and unloved and left to fend for himself.*
(COCA: Elizabeth Thornton, *The Worldly Window*, New York: Zebra Books, 1990)

(7) *I mean, if you show me a woman on drugs and I'll show you the guy who got her started on it.* ***No woman*** *actively seeks out drugs. It's usually introduced to her by somebody else because a woman has an innate sense to protect themselves because they're more valuable in a reproductive sense than a man is. So, it's always the man who takes the risk and then unfortunately drags women into his world.*
(COCA: Fox Five, February 13 2012)

(8) *"Come on, brother," he said, "you have to smoke this now."* ***No man*** *likes to hear the words "have to" from another man. But he was my only link to pot and the other drugs I was also sometimes taking, so I gave in and smoked.*
(COCA: Richard Burgin, *The Dealer*, Antioch Review, Vol. 67 Issue 4, 2009)

A manipulation of example (8) shows that the plural form is unacceptable:

(8') *No men like to hear the words "have to" from another man.*

Furthermore, a shift in the position of the negation in the sentence highlights the generic character of the noun determined with *no* and the singular form:

(6') *A child should never be alone or unloved and left to fend for himself.*
(7') *A woman does not actively seek out drugs.*
(8") *A man does not like to hear the words to hear the words "have to" from another man.*

The determiner *a* here does not refer to any particular and situated occurrence, which shows that it is the generic reference that is dealt with. In addition, a closer look at the co-text, highlights the fact that the sequence *no + singular noun* is most compatible with a modal auxiliary referring to norms or principles, such as *should* in (6), or with the simple present forms, as exemplified by *seeks* and *likes* in (7) and (8).

In fact, the predicates *be alone*, *like to hear the words 'have to'*, and *seek out drugs* refer to criteria which, according to the subjective origins of those discourses, do not pertain to the notion /CHILD/, /MAN/, and /WOMAN/. Those predicates cannot be considered as constitutive properties of those notions.

*No + singular noun* thus indicates the absence of any element with the properties of the notion referred to (the notion /CHILD/ or /MAN/ or /WOMAN/), which means that this absence is considered on an unlimited class of elements.

These examples should be contrasted to examples (9), (10), (11) and (12):

(9) *The review of athletic director job announcements provides insight into how women are being excluded from administrative positions in interscholastic athletics. [...]. Since* ***no women*** *in Texas are head football coaches and very few, if any, are assistant football coaches, it becomes analogous to sex-segregated newspaper job announcements that have been deemed to promote unallowable employment discrimination due to sex.*
(COCA: Miller, John J. / Whisenant, Warren A. / Pedersen, Paul M., "The Communication of Opportunities and Barriers to Prospective Applicants: An Analysis of Interscholastic Athletic Administrative Job Announcements", Physical Educator, Vol. 64 Issue 2, 2007)

(10) *But here's the big question, if I can focus us, the American people have a right to know that there's been a shift in our, in our food supply for apple juice. I'm a big supporter of apple juice. I have talked about it on my show. I continue to give it to my kids. I wanna be clear about this,* ***no children*** *are dying from acute toxicity from arsenic. The levels we're talking about are much lower than that. My concerns are about long-term effects, over years of kids getting arsenic levels that are higher than we can tolerate.*
(ABC-GMA, "Apple Juice Episode Under Fire; Experts refute arsenic test, 2005)

(11) *More men than women also indicated that they believed they were gifted in high school but not gifted now (17% of men and 10% of women).* ***No men*** *indicated that they did not believe they were gifted in high school but believed they were gifted now; however, 23% of women fit this category.*
(COCA: Kristin M.Perrone, "Self-Perception of Gifts and Talents Among Adults in a Longitudinal Study of Academically Talented High-School Graduates", Roeper Review, Vol. 29 Issue 4, 2007)

(12) *A sign to Delacroix Island was gone, along with the rest of the town. It had said, appropriately, "Fin de la Tierra", meaning End of the Earth. Bicycles and washing machines clung to trees.* ***No dogs*** *hungrily roamed the streets, as in other towns closer to New Orleans. There was no life, except for the sea gulls cawing over the wreckage of the fishing industry and the homes where the fishermen had lived.*
(COCA: Chicago Sun Times, "Like their Islenos ancestors, storms survivors persevere", 2005)

Now, once again with those utterances, shifting the negation to the verbal phrase highlights the specificity of the marker *no + plural*:

(12') *The dogs did not hungrily roam the streets.*
(10') *The children are not dying from acute toxicity from arsenic.*
(11') *The men did not indicate that they did not believe they were gifted in high school.*

This manipulation shows in what way the sequence *no + plural* is different from *no + singular*. The compulsory use of the definite article *the* before the plural nouns indicates that the utterances refer to a specific lot of elements, defined in the situation. It should also be noticed that in the above examples aspects and tenses also refer to specific situations: *be + V-ing* in (10) and the preterit in (11) and (12).

In (9), (10), (11) and (12), the enunciator constructs the absence of any occurrence belonging to a class of occurrences and locates this absence relative to a definite situation, which is the space-time frame referred to in the context. The speaker posits the absence of specific occurrences in a given situation. As a result, the absence is considered relative to a particular class of elements within a specific space-time frame. Beyond the potential differences between these elements (beyond heterogeneity), all of them can eventually be regarded as equivalent since they all negate the predicate.

In other words, the events or situations referred to are situated within a specific space-time frame and the sequence *no + plural noun* indicates the absence of all the elements belonging in this particular space-time frame.

Example (13) is yet another eloquent illustration of this:

(13) *"Football may be the last bastion of male domination," said Chuck Neinas, executive director of the College Football Association, when he addressed CFA coaches and athletic directors May 29. "Just as there are* ***no men*** *who are mothers,* ***no woman*** *has played college football."*
(Ben Brown, "Equity: Football sees status up in the air", USA Today, 1992)

Here, the sentence (13') ?**Just as there is* ***no man*** *who is a mother*, (…) would be disallowed whereas (13")? ***no women*** *have played college football* is more acceptable.

It is worth noticing that the sequence in the plural is introduced with a predication of existence, delimitating the negation to a specific situation, and the singular fits with a present perfect form, compatible here with the expression of genericity.

Deleting the predication of existence would allow the use of the singular and render (13') acceptable:

(13a) *Just as* ***no man*** *is a mother,* […]

Changing the present perfect in the second part of the sentence into a preterit and adding a time or space locator would have the same effect on (13"):

(13b) ***no women*** *played college football at that time/in our school.*

From the speaker's point of view, the action *play football* does not fit with the notion /WOMAN/, it is not one of its constitutive properties. On the contrary the sequence *no men* does not rule out potential differences between individual men (in other words between the elements of the class), even if the common feature is that a man cannot be a mother.

The plural form introduces heterogeneity. As a result, any element of the class may potentially be distinguished from any other element of the same class. Nevertheless, in the end, all the occurrences are set on a par since none of them is valid for the predicate.

As a consequence, we hypothesize that the pattern *no + N-Ø* exhibits the core properties of the notion referred to while the pattern *no + N-s* retains potential differential properties, as epitomized in example (14):

(14) *And* ***no policemen*** *he noticed with gratification, neither Kenyan, nor British. Had Bernard Pellegrin's potions worked their magic? Whisper who dares.*
(*The Constant Gardener*, John le Carré, 2001)

In this example, there is an explicit reference to a finite lot of elements; the mention of their nationalities suggests that the character, the subjective point of view in this passage, was expecting specific instances of policemen.

However, the manipulations on example (13) tend to prove that there is some independence between the utterance and what it refers to. Far from depending only on an actual pragmatic feature, as M. Swan (1995) suggested with *no wife/no children*, the use of *no +N-Ø* versus *no +N-s* is related to a subjective construction of meaning. Both may refer to the same fact but considered from a different point of view:

(13a) *Just as* ***no man*** *is a mother,* […] would mean that the utterance would be located relative to a subjective source, asserting their own definition of the notion MAN.
(13) *there are* ***no men*** *who are mothers* is, on the contrary, located relative to an existential frame and does not refer to the notional definition.

This hypothesis may help us account for the utterances exhibiting the co-occurrence of the two patterns under study as seen in example (1) which introduces this article:

(1) *MADIGAN: I don't work for the school district, but I do know this,* ***no teacher*** *has complained to the teacher's union, and the video was shown over a month ago. I find it hard to believe that the board had any real issue with it. HILL: Board members are expected to take up this issue of Mitchell's future, Tuesday night. (END-VIDEOTAPE) S-O'BRIEN: That story from Reporter Darlene Hill, from CNN affiliate, WFLD, reporting. Interesting there, huh? M-O'BRIEN: Yes, I can see how the humor might be misconstrued. S-O'BRIEN: But* ***no teachers*** *complained. And the board, so far, is the only people who have problems with it. Maybe there's something else at work. M-O'BRIEN: There's a fair amount of politics in schools, you might say. You can bet on that, right? That's a sure bet. SERWER: Yes, you sure can.*
(COCA: CNN-AM, 2006)

In this excerpt, the aspectual determination on the verb forms, the present perfect as opposed to the past-simple or preterit, is a most important feature. The present perfect induces the speaker to interpret past events together with their present results: with *no teacher has complained* the speaker rules out the existence in the present of any complaint. The present perfect highlights the relationship between the speaker and the event referred to. Therefore, with the present perfect and the singular form in *no teacher has complained*, the speaker means "not a single person with the qualities /TEACHER/" and thus ascribes a kind of prototypical dimension to the element *teacher*. This high degree of abstraction has to be related to the speaker's viewpoint and subjectivity.

On the contrary, the preterit indicates that the event does not depend on any particular point of view. In *no teachers complained*, the preterit gives information in a factual and objective way. The information given is not new but focuses on the fact itself, giving it a time and space delimitation, all the more so as what actually happened stands in contrast to what was expected (as shown by *but*). The teachers referred to are those working for the school district. Consequently, with the combination of the preterit and the plural form in *no teachers complained*, the speaker places himself from the same standpoint as that of the addressee.

The difference between the two sequences in the utterance under study is enhanced by the presence of *I do know this* just before *no teacher*, highlighting a subjective and evaluative dimension, as opposed to *you might say, you can bet on* ***that*** in the vicinity of *no teachers*.

## Conclusion

The study of the alternation between the *no + N-Ø* and *no + N-s* patterns led to an analysis of the complexity of design pertaining to the reference to absence, which has to be described and elucidated if significant results are to be achieved.

We have dealt with processes, with nothing at our disposal but the end products—in other words the material traces in the texts examined. As a result, an attempt was made to find solutions in the outputs themselves. The reference to absence was highlighted and made more precise with the notions of preconstruction, generic reference and specific reference. These reflect very subtle differences in meaning that have to be brought into clear focus and accounted for. We are thus led to conclude that these phenomena map the parameters of meaning and should be analyzed at the interface between syntactic, semantic and discursive features.

We were led to conclude that the singular and the plural indicate specific linguistic operations regarding negation:

- the construction of a class of elements with *no* + *N-s* refers to the absence of the whole class whereas *no* + *N-Ø* refers to the absence of the properties of the notion dealt with;
- the negation of the whole class of elements is located relative to a specific situation whereas the predication of the absence of properties of the notion dealt with keeps the trace of a subjective evaluation.

Finally, while this study provides a glimpse into the potential of this kind of approach, it still has to be tested with other negative patterns.

## References

*Bouscaren, J. & al.* (1997). Cahiers de recherche en grammaire anglaise, T. 7, *La composante qualitative – Déterminants et anaphoriques* [Qualitative features in determiners and anaphora], Université Paris VII, Gap: Ophrys.

*Culioli, A.* (1990). Representation, referential processes and regulation. Language activity as form production and recognition. In *Pour une linguistique de l'énonciation, Opérations et représentations, Tome I*, Collection HDL, pp. 177–213. Gap: Ophrys.

*Culioli, A.* (1995). Cognition and Representation in Linguistic Theory. In M. Liddle (Ed.) *Current Issues in Linguistic Theory* (vol. 112). Amsterdam/Philadelphia: John Benjamins.

*Culioli, A.* (1997). A propos de la notion [About the notion]. In *La Notion*, Collection HDL, pp. 9–24, Gap: Ophrys.

*Gilbert, E.* (1999). *SOME* et la construction d'une occurrence [*SOME* and the construction of occurrences], *CYCNOS*; vol. 16–2, Nice.

*Gilbert, E. & Le Querler, N.* (1998). *La référence 1 – Statuts et processus* [Reference: status and process], Cerlico, Rennes: Presses Universitaires de Rennes.

*Gournay, L.* (2007). *linguistique.wifeo.com/documents/HDR-Gournay.pdf*

*Guillemin-Flescher, J.* (1981). *Syntaxe comparée du français et de l'anglais – problèmes de traduction* [A comparative Syntax of French and English – Problems in translation], Ophrys.

*Huddleston, R. & Pullum, G. K.* (2002). *The Cambridge Grammar of the English Language, Cambridge: Cambridge University Press.*

*Odin, H.* (1998). Les traductions de *any* dans le passage de l'anglais au français [Translating ANY from English to French]. In *Linguistique Contrastive et Traduction* T4, J. Guillemin-Flescher (Ed.), Gap: Ophrys.

*Quirk, R., Greenbaum, S., Leech, G. & Svartvik, J.* (1985). *A Comprehensive Grammar of the English Language*, London: Longman.

*Swan, M.* (1995). *Practical English Usage*, Second Edition, Oxford University Press.

*Rivière, C.* (1997). Qualités méconnues de *any* [Unveiling the properties of ANY] in J. Bouscaren (Ed.), Cahiers de recherche T.7, *La composante qualitative: Déterminants et anaphoriques*, pp. 93–108, Gap: Ophrys.

Dylan Glynn
*University of Paris VIII, France*

# The socio-cultural conceptualisation of FEMININITY: corpus evidence for cognitive models

**Abstract:** This study examines the possibility of extending Multifactorial Usage-Feature Analysis (Profile-Based Approach) to describe abstract conceptual structures such as those identified in Idealised Cognitive Model (ICM) research. The approach is argued to resolve two methodological limitations of the analytical framework of ICM. These limitations can be described as (i) a lack of means for identifying social variation in the structure posited and (ii) a lack of means for falsifying the structures identified with the framework. Multifactorial Usage-Feature Analysis is corpus-driven and quantified, permitting a multi-dimensional picture of the models that accounts for social variation as well as falsification through repeat analysis. The study focuses on the concept of FEMININITY. Instead of limiting the analysis to metaphoric structure, it takes a keyword lexical approach. The data are synchronic and restricted to a specific genre / register of American English.

Keywords: Idealised Cognitive Models, Multifactorial Usage-Feature Analysis (Behavioural Profile Approach), corpus linguistics, Cognitive Linguistics, femininity.

## 1. Introduction: Usage-Based Cognitive Models and FEMININITY

The Idealised Cognitive Model, first proposed by Lakoff (1987), has been systematically demonstrated to be a powerful and versatile tool for the description of conceptual structure as encoded in language. However, two fundamental concerns regarding its empirical reliability remain to be considered. Firstly, the very nature of the models are 'idealised'. This means that they are extremely abstract generalisations that offer no information as to socio-cultural variation. In other words, each language community, or culture, is treated as a single block, with the caveat that not every individual shares all the facets of a single Cognitive Model. It is for this reason they are termed "idealised". Since Cognitive Linguistics assumes a usage-based model of language, variation is held to be basic to language structure, and therefore in turn, culture. Current methods for the identification of cognitive models omit any possibility for capturing such variation. Moreover, and perhaps more importantly, this very point makes the falsification of the models extremely difficult. Since counter evidence observed may represent variation in the langauge community, it is not possible to disprove the proposal of any given model. Methodologically, this is a serious weakness that brings into question the value of the results obtained using this framework.

This study seeks to amend this situation by proposing a method of analysis that will integrate social variation into its results and permit the falsification of those results. Extending Multifactorial Usage-Feature Analysis to Cognitive Models has been attempted before. Glynn (2007, 2014e, Glynn 2014f), Sten & Glynn (2011), Nordmark & Glynn (2012) and Krawczak (2014b; 2014c; this volume) all represent attempts at extending the method in this manner. However, these previous studies, with the exception of Sten & Glynn (2011) and Glynn (2014e) restricted themselves to emotion concepts where the rich metaphoric structure was the focus of attention. Sten & Glynn (2011) and Glynn (2014e) apply the method to an abstract concept where metaphoric structuring is less important, like with FEMININITY, but these studies restrict the analysis to a single lexeme. This study attempts to apply the method to non-metaphoric language across a set of near-synonyms, or keywords.

The concept of FEMININITY is obviously a socially pertinent concept. In cultural studies, critical discourse analysis just as in sociology and social anthropology, the conceptualisation and representation of gender is a cornerstone of research. Moreover, growing research in cross-cultural pragmatics (Blum-Kulka *et al.* 1989, Wierzbicka 1991) and corpus-based Pragmatics will be directly informed by the successful application of Multifactorial Usage-Feature analysis to this concept. Although the current study is effectively a proof-of-principle study, the results represent an advancement in the quantitative description of socio-culturally sensitive concepts using corpus data.

## 2. Method and data

### 2.1 Multifactorial Usage-Feature Analysis and Key Word Analysis

The investigation employs Multifactorial Usage-Feature Analysis or the Profile-Based Approach (Dirven *et al.* 1982; Geeraerts *et al.* 1994; Gries 2003). This methodology was developed amid early research in Cognitive Linguistics designed to permit corpus-driven quantitative analysis of purely semantic phenomena. Moreover, computational linguistics has recently adopted the method. In Computational linguistics, it is termed Sentiment Analysis or Opinion Mining (Wiebe *et al.*). The method is straightforward and is based on qualitative analysis (manual annotation) of large numbers of contextualised occurrences of a given linguistic phenomenon. These occurrences take the form of a sample extracted from a corpus, either based on a formal string (semasiological) or on a manually identified linguistic function or concept (onomasiological analysis). The analysis / manual annotation comprises the systematic application of a set of pre-determined conceptual-functional categories or usage-features to each of the occurrences / examples. The occurrences are each tagged (annotated) for these features. This produces a large set of 'meta-data' concerning the use, or behaviour, of the linguistic phenomenon under investigation. For a more detailed description of the method and its strengths and weaknesses, see Glynn (2010b; 2014b; 2014c). Within Cognitive Linguistics, a few examples of its use would include Dirven *et al.* (1982), Geeraerts *et al.* (1994), Gries (1999; 2006), Divjak (2006; 2010), Divjak & Gries (2006), Gries & Divjak (2009), Glynn (2007; 2008; 2009; 2010a; 2014a), Janda & Solovyev (2009), Krawczak & Glynn (2011), Krawczak & Kokorniak (2012), and Krawczak (2014a). There are also three edited volumes devoted, in part, to developing the method, Gries & Stefanowitsch (2006), Glynn & Fischer (2010), Glynn & Robinson (2014).

In Cognitive Linguistics, keyword analysis is used to operationalise the study of conceptual structure (Wierzbicka 1985; Kövecses 1986; Lakoff 1987; Vorkachev 2007; Bartmiński 2009). This method assumes that the lexical semantics of representative 'keywords' can be employed as an index of the language community's conceptual structure. The keyword principle can be adopted *mutatis mutandis* and integrated into Multifactorial Usage-Feature Analysis. The principle is simple: take the keywords traditionalised analysed as abstract lexical categories using introspection and use them as keywords in the retrieval of natural usage examples in corpora. In this way, combining Wierzbicka-style keyword analysis and Multifactorial Usage-Feature Analysis is straightforward.

## 2.2 Lexemes and corpus sample

In this study, the choice of keywords was operationalised in terms of frequency. An exhaustive list of adjectives denoting FEMININE were taken from thesauri. Using the Corpus of Contemporary American English (Davies 2008), the six most frequent lexemes were then determined. These lexemes were *motherly*, *girly*, *girlish*, *feminine*, *womanish* and *woman*. The data were taken from the American component of the LiveJournal Corpus (Speelman & Glynn 2005). This corpus is stylistically homogenous, being restricted to personal diaries, written largely by young British and American students. For each lexeme, 200 examples for each form were extracted, each with a large context of 200 words left and right. Occurrences that were quotations, adverbs or from the titles of diary entries were omitted. The frequencies of the lexemes and lemmata after cleaning are presented in table 1.

*Table 1: Lexeme and lemma frequency of* FEMININE *keywords*

| Lexeme | Frequency | Lemma | Frequency |
|---|---|---|---|
| *feminine* | 201 | FEMININE | 201 |
| *girlish* | 139 | | |
| *girly* | 69 | GIRLY | 208 |
| *motherly* | 120 | MOTHERLY | 120 |
| *womanish* | 80 | | |
| *womanly* | 69 | WOMANLY | 149 |

## 2.3 Analysis. Manual annotation of sample

The usage-feature analysis in this study examines 4 different dimensions. Each dimension, or factor, is designed to come together to offer a picture of the representation, and arguably the conceptualisation, of FEMININITY. The factors include Topic of Discourse, Referent Type, Referent Gender, and Evaluation. Each factor is explained and exemplified below.

### *2.3.1 Topic of Discourse*

The topic of discourse is a crucial factor in this analysis. In language and culture, gender is associated with certain roles and social domains. The annotation of Topic of Discourse permits the identification of the correlation between the conceptualisation of FEMININITY and various gender roles.

(1) Emotion & psychology
Ask a parent, neighbor, sibling, or close friend whom you trust to keep them for a few hours while you regain your *motherly* strength.

(2) Entertainment & fame
the car was made famous among the female species thanks to the constant promotional effort by the most girly and popular hotel heiress, Paris Hilton.

(3) Family & love
Ray has always been ready and willing to learn and help, and trusts my *motherly* knowledge when it comes to things like not eating orangey baby food.

(4) Gender & stereotype: because the *feminine* dimension is something of a problem for the difference between the sexes.

(5) Health & appearance
I would love to develop a very *feminine*/traditional waistline through belly dancing.

(6) Science, religion & politics
A year ago the New York Times launched a special "*feminine* section" on women changing the world, and the Dalai Lama flattered many of us, when he declared that the Western Woman will save the world.

(7) Fashion & decor
They are very cute, very *womanly*. Flower motif was obviously dominant and it was good.

(8) Behaviour & attitude
I was much more afraid that she would have given him the *motherly* finger wag and then would have gone on to praise Vick for "turning the corner," regaining his equilibrium, etc.

(9) Art & culture
The line has all of our core aesthetic – clean details, rich prints, always a masculine meets *feminine* moment, and details without being frilly or overly girly in any way.

### *2.3.2 Referent Type*

Referent Type is a basic semantic factor which indicates the referent of the adjective. It is designed to answer the questions: Are the feminine adjectives being used to talk about people or things and, if so, what kinds or things and what kinds of people. The annotation of this factor can be highly subjective. For this reason, coarse grain categories are used. In the confirmatory analysis below, the categories are further conflated to make a binary distinction between animate and inanimate.

(10) Inanimate Abstract
*feminine* intuition stuff sells magazines, but in real life it's still a fairy tale.

(11) Inanimate Concrete
*Feminine* tattoos are generally smaller and far prettier than those for men.

(12) Inanimate Activity
I've never known any male who thought it was *womanly* or effeminate to cook.

(13) Inanimate SoA (state-of-affairs)
For example, in the film London to Brighton, a prostitute is shown in a very *motherly*, caring way and..

(14) Inanimate Event
"If you don't graduate from college, you won't get a good job," I responded with a tinge of panic and *motherly* concern.

(15) Human Specific
She's more stylized, she's more *feminine*, she's more a woman as she dances this time around.

(16) Human Generic
As for me, *girly* girl is likely to be a girl that wear skirt, fancy headband, sweet pumps, "kawaii" accessories, other pink stuffs, and love ballet.

### 2.3.3 *Referent Gender*

The gender of the referent is obviously crucial in gauging the use of the adjectives. Four categories were needed to account for the data. Masculine and feminine referents, but also unknown and mixed gender referents had to be allowed for. Finally, a category of 'non applicable' was used for inanimate referents.

(17) Female
The woman had a very kind motherly tone in her voice, which reminded her suddenly of her mother that had passed away 5 years ago.

(18) Male
He acts *womanish* toward Hearn, scorned and jealous, lashing out needfully.

(19) Unknown
For a minute, let's forget about *girly* and focus on non-manly. I am in the process of building a 2nd desktop.

(20) Mixed
The fact that so many White's accept these lies shows how *womanish* and effete the race has become.

(21) Non Applicable
It seems to me there should be a widely accepted vision of what *feminine* power should look like, and…

### 2.3.4 *Evaluation*

The category of evaluation is highly subjective. The annotation is operationalised by assuming that the use is neutral unless there is clear textual indication to the contrary. Typically the use of adverbs and adjectives were the criterion used to determine either positive or negative evaluation. This rule is not, however, categorical and, for certain occurrences, encyclopaedic semantics were used, as in example (24).

(22) Neutral
One trend that appears to be emerging in the collections this Fashion Week is the blend of the masculine and the *feminine*.

(23) Positive
The New Look bomb astonished with its flamboyant colors and wonderful *womanly*, flower-like silhouettes that altered style codes and brought desire back to life.

(24) Negative
If it were not for the bloodshed there, his rantings on TV this morning in a womanly voice would have been comical.

## 3. Results

### 3.1 Mapping the lexical semantics

Due to practical limitations, this study will be restricted to an analysis at the level of lemma. The first step was to identify correlations between Topic of Discourse and the different lemmata. Figure 1 presents the results of a multiple correspondence analysis that identifies the correlations between Lemma, Topic of Discourse, Referent Type and Evaluation.

Biplots, such that that in Figure 1, represent relative degrees of associations (positive correlation) and disassociation (negative correlation) through the proximity of the data points in the two-dimensional plane. As such, each biplot is a two-dimension representation of multidimensional space, where the correlation of all of the factors (features) are calculated simultaneously. The plots also represent distinctiveness in correlation by placing the data points closer to or further from the *x*- and *y*-axes; the closer to the axis, the less distinctive it is along that dimension. It follows that data points close to the centre are the least distinctive in usage across both dimensions. Furthermore, the contribution, or relative importance, of each of the data points is represented by the size of the data point; smaller points contributing less to explaining the behaviour of the data, larger points being more important. For a more detailed explanation on how correspondence analysis operates and how to interpret the plots, see Glynn (2014b).

*Figure 1. Lexical semantic structure: Multiple correspondence analysis Correlations between Lemma, Topic of Discourse, Referent and Evaluation.*

In figure 1, we see the dispersion of the usage-features which characterise the use of each of the lemmata. Focusing, at first, on the gender of the referent, we see that Female referent lies directly between the lemmata GIRLY and FEMININE. This means that these two lemmata are equally associated with this referent gender. Moving across the plot, we see that Gender N.A., which refers to inanimate referents, lies between FEMININE and MOTHER. Finally, on the right of the plot we find WOMANLY distinctly associated with Male referents.

However, it is important to note that these associations are relative to both the Topic of Discourse and the Referent Type. In other words, is not simply that these gender referents are associated with these lexemes, but that these genders are associated with these referents, when treated in the context of the other factors. Therefore, on the right, where we see the features 'negative' Evaluation, 'specific human' Referent and the Topic of Discourse 'behaviour and attitude', it is the combination of these features that represents the characteristic pattern, rather than any single association.

Moving across to the left, we see a cluster of features that lie between GIRLY and FEMININE. We can assume that these features are equally associated with each lemma but that a specific feature is disguising them. It appears that although both GIRLY and FEMININE are equally associated with 'female' Gender Referents, the Topics of Discourse 'fashion & decor' 'art & culture', 'health & appearance', as well as 'inanimate concrete' Referents, they a distinguished by one important difference: GIRLY is distinctly associated with 'generic human' Referents in contrast to feminine which is being drawn up away from the cluster towards 'inanimate abstract' Referents. At the top of the plot, MOTHER is quite distinct in its use. Although it lies between FEMININE and WOMANLY for most of the usage features, it appears distinctly associated with the Topics of Discourse 'gender & stereotype' and 'emotion & psychology'.

None of the associated features identified are surprising, indeed, they are all intuitively sound. This tells us that the overall behaviour of the lemmata matches what one would expect given introspective consideration of their use. The next step it to remove the lexemes from the analysis in an attempt to identify the structure of the concept *per se*, rather than the onomasiological structure of the lexemes.

## 3.2 Mapping the conceptual structure

Having established the onomasiological structuring of the concept by charting the use of the lemmata relative to the range of usage-feature under investigation, we can now examine the patterning of those features without the lexemes to structure them. Before we interpret the results, we need to examine the reliability of the depiction, that is how well the correspondence analysis been able to take a highly complex multidimensional structure and represent it in two-dimensions. Table 2 includes the scree plot of the dimension reduction. The plot represents the first two dimensions listed in the table, together accurately representing 63.2% of the variation, or 'inertia'. For such a complex analysis, this figure is reasonable. Of more concern is the fact that there is no clear 'elbow' in the scree plot. In other words, as dimensions are added, at no specific point does the accuracy of representation cease to improve substantially. In other words, there is a gradual decline in the contribution to explaining variation as we add dimensions but at no point can we say a given number of dimensions is sufficient to explain the underlying structure of the data. A possible interpretation of the gradual (as opposed to abrupt) decline in inertia is that the analysis is struggling to represent the complexity of the situation in two-dimensions and we need to be cautious in our interpretations.

*Table 2: Conceptual Structure: Multiple correspondence analysis, inertias (eigenvalues)*

| Dim. | Eigenvalues | % of Inertia | Cumulative % | Scree Plot |
|---|---|---|---|---|
| 1 | 0.132793 | 44.9 | 44.9 | ************************* |
| 2 | 0.054108 | 18.3 | 63.2 | ********** |
| 3 | 0.029802 | 10.1 | 73.3 | ****** |
| 4 | 0.015083 | 5.1 | 78.4 | *** |
| 5 | 0.008478 | 2.9 | 81.2 | ** |
| 6 | 0.004128 | 1.4 | 82.6 | * |
| 7 | 0.002526 | 0.9 | 83.5 | * |
| 8 | 0.001658 | 0.6 | 84.0 | |
| 9 | 0.000589 | 0.2 | 84.2 | |
| 10 | 0.000146 | 0.0 | 84.3 | |
| 11 | 00000000 | 0.0 | 84.3 | |

Since caution is needed in the interpretation, Table 3 offers a break down of each of the features in the analysis and the quality of their representation as well as their contribution to explaining the behaviour of the data along the two axes. Although no absolute rule is possible, any quality scores that are less than 500 in 'quality' should be treated with caution (Greenacre 2007). Particularly problematic are the Topics of Discourse 'art & culture' and 'science & politics', and the Referent Types 'human generic', 'human specific', and 'inanimate abstract'. The depiction of the Evaluation features are all extremely reliable.

*Table 3: Conceptual Structure: Multiple correspondence analysis, quality and contribution*

| Feature | | Quality | x - Axis | Contribution | y - Axis | Contribution |
|---|---|---|---|---|---|---|
| ToD.: | art & culture | 111 | -128 | 1 | -231 | 7 |
| ToD.: | behaviour & attitude | 878 | 757 | 185 | -183 | 26 |
| ToD.: | emotion & psychology | 584 | 262 | 9 | 499 | 83 |
| ToD: | entertainment & fame | 255 | -118 | 1 | 246 | 10 |
| ToD.: | family & love | 530 | 135 | 3 | 516 | 113 |
| ToD.: | fashion & decor | 674 | -501 | 132 | -78 | 8 |
| ToD.: | gender & stereotype | 462 | 262 | 27 | 170 | 28 |
| ToD.: | health & appearance | 407 | -162 | 19 | -228 | 92 |
| ToD.: | science & politics | 191 | -105 | 1 | 309 | 27 |
| Ref.: | FEM + Inanimate Abstract | 595 | -10 | 0 | 309 | 117 |
| Ref.: | FEM + Inanimate Concrete | 816 | -505 | 95 | -178 | 29 |
| Ref.: | FEM + Inanim. Event-Activity | 557 | 499 | 25 | 446 | 49 |
| Ref.: | FEM Human Generic | 591 | 138 | 1 | -330 | 17 |
| Ref.: | FEM Human Specific | 202 | 57 | 1 | -118 | 7 |
| Ref.: | FEM Inanimate SoA | 562 | 260 | 5 | 369 | 27 |
| Ref.: | Human Generic | 175 | -101 | 1 | -326 | 25 |
| Ref.: | Inanimate Abstract | 221 | -165 | 10 | 187 | 31 |
| Ref.: | Inanimate Concrete | 516 | -433 | 46 | -274 | 45 |
| Ref.: | Inanimate Event-Activity | 550 | 511 | 24 | 8 | 0 |
| Ref.: | MALE + Inanimate Abstract | 754 | 586 | 41 | -40 | 0 |
| Ref.: | MALE + Inanimate Concrete | 560 | 431 | 10 | -182 | 4 |
| Ref.: | MALE Human | 756 | 623 | 59 | -377 | 53 |
| Evaluation: | Negative | 930 | 679 | 209 | -257 | 74 |
| Evaluation: | Neutral | 870 | -94 | 13 | 147 | 77 |
| Evaluation: | Positive | 902 | -285 | 49 | -162 | 39 |

A note of explanation is due here concerning the concatenation of the Referent Gender and Referent Type. In order to minimise the dimensionality of the analysis, the Gender of the referent and the Referent Type have been combined. This means that, in some instances, we have a combination of a gender referent and an inanimate referent together. This is due to the fact that there are many uses where the adjective describes an inanimate feature associated with a human, such as, for example, *he has a feminine voice*. In such instances, although the actual referent is inanimate ('voice'), the fact that the voice belongs to a man is the most relevant characteristic.

*Figure 2. Conceptual Structure: Multiple correspondence analysis Correlations between Topic of Discourse, Referent Type and Evaluation.*

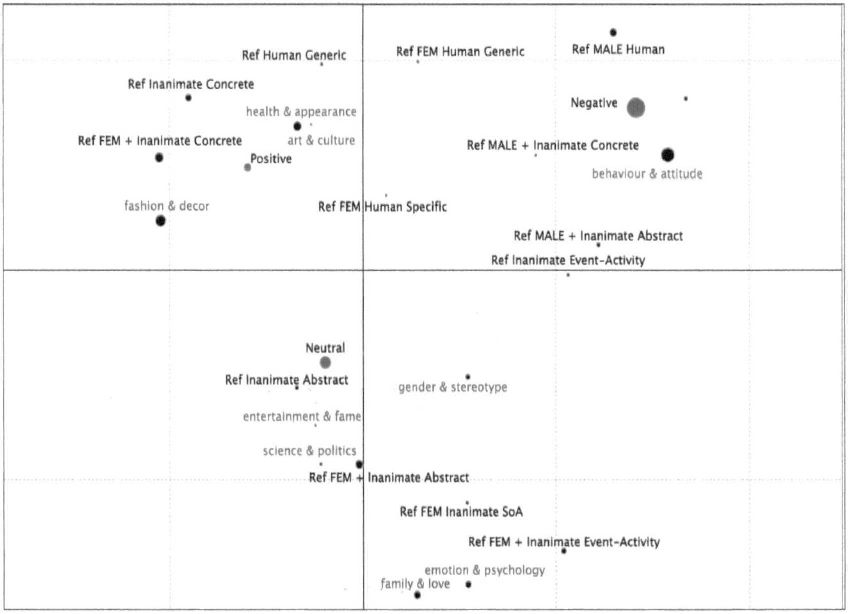


Figure 2 presents the results of a multiple correspondence analysis that includes all the usage-features but not the lexemes. The data points appear to cluster into three broad groups. These can be understood as sub-categories of the concept of FEMININITY. In the top-right, dominated by 'negative' Evaluation, we have the clustering of 'male' Referent Gender as well as the Topic of Discourse of 'behaviour & attitude'. This cluster mirrors what we saw above and seems to be both intuitively sound and clear: FEMININITY as a behaviour, especially when associated with men, is negative.

To the left of the plot, another relatively clear cluster is dominated by 'positive' Evaluation. Here the Topics of Discourse are 'fashion & decor', 'art & culture', and 'health & appearance'. The Gender of the referents is a more complex picture than for the right-hand cluster. 'Female' Referents and 'inanimate objects' as Referents are clearly associated with this clustering, but specific 'female' Referents actually lie between this cluster and the 'negative' Evaluation – 'male' Gender cluster. This would suggest that when the Referent Gender is 'female' and the focus of the discussion is behaviour, 'negative' evaluation is common. Further research would be need to verify this, but it is an astounding result.

The bottom cluster is broad, stretching across the $x$-axis and could be argued to pre-resent two sub-clusters. However, given the homogenous nature of the features clustered, interpreting this as a single pattern appears most reasonable. The cluster is dominated by 'neutral' Evaluation and a range of 'inanimate' and 'female' Referent Types. It would seem that states of affairs, activities and events, associated with women are evaluated as neutral and that such a conceptualisation is associated with the Topics of Discourse of 'entertainment & fame', 'science & politics', 'family & love', and 'emotion & psychology'. The Topic of discourse 'gender and stereotype' is also grouped with this cluster but is drawn up towards the right top of the plot and the cluster of male referents and negative evaluation of behaviour.

## 3.3 Confirmatory analysis of conceptual results

Although we have considered the stability of the results above and have identified structures in the sample that we can, with confidence, interpret as conceptual structures, correspondence analysis offers no information about the probability that the correlations and patterns identified are representative of the language more generally (beyond our sample). In order to obtain information about the significance of the identified structures, we need to turn to confirmatory analysis. Due to the need for a larger data set relative to the number of factors under investigation, the results of the confirmatory analysis are necessarily less broad than that of those obtained through correspondence analysis. Moreover, just because one does not obtain a significant correlation in the confirmatory analysis, where we do see a correlation in the correspondence analysis, does not indicate a falsification of the previous result, merely that we have not been able to confirm that result. Data sparseness is just as likely a cause of non-confirmation as a misleading result in the previous section.

We employ loglinear analysis in the confirmatory stage (Agresti 2013). This method is essentially a large number of Chi-squared tests that seek to ascertain if the positive and negative correlations observed above are significant. In Figure 3, the results of the loglinear analysis are presented in the format of a mosaic plot.

*Figure 3. Conceptual Structure: Mosaic plot of loglinear analysis. Significant positive and negative multidimensional correlations of Evaluation, Gender, Animacy and Topic of Discourse.*[1]

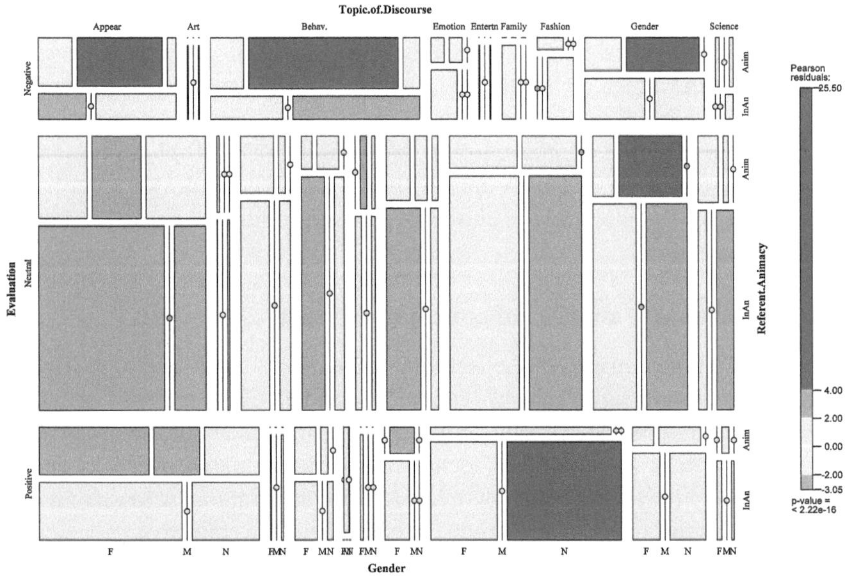


The visual representation of the results of the loglinear analysis are detailed and somewhat difficult to interpret. They plot significant positive and negative correlation between 4 factors simultaneously. Therefore, each box in the plot represents the co-occurrence of 4 different features. The blue boxes represent a significant positive correlation (association) and the red boxes a significant negative correlation (disassociation). The darkness of the colour, blue or red, indicates relative effect size, darker representing strong effect. The size of the box is determined by relative frequency and grey boxes mean that the association is not statistically significant.

The most striking result is across the top row where we see significant associations between 'negative' Evaluation, the Topics of Discourse 'appearance', 'behaviour' and 'gender' and 'animate' 'male' referents. Although this is completely in line with what the correspondence analyses revealed, it is here confirmed as significant.

1 Interpreting Figure 3 in grey scale is not possible. The colour version of the figure can be found at www.dsglynn.univ-paris8.fr/publications.html.

Again confirming what we saw above, significant positive correlation between 'neutral' Evaluation, 'inanimate' & 'female' referents and the Topics of Discourse of 'appearance', 'behaviour' and 'fashion' is revealed. One of the largest significant correlations is between 'positive' Evaluation, 'inanimate' Referent and the Topic of Discourse of 'fashion'. Two significant negative correlations are observed. Firstly, 'female' & 'inanimate' referents do not co-occur with 'negative' Evaluation in the Topic of Discourse of 'appearance'. Secondly, 'neutral' Evaluation does not combine with 'inanimate' Referents also for the Topic of Discourse of 'appearance'. Save the this latter negative correlation, all the observed associations align with what we have observed in the correspondence analysis and match an intuitive picture of the conceptualisation of FEMININITY.

## 4. Summary

The study revealed systematic patterns in the usage of the lexemes denoting FEMININITY. Without lexical or metaphoric structuring, a combination of Referent, Evaluation and Topic of Discourse revealed three subcategories of the representation of the concept amongst young speakers in personal diaries. Two of these multidimensional subcategories were confirmed and demonstrated to be statistically significant.

Subcategory 1 – statistically confirmed
Referent: Male
Evaluation: Negative
ToD: Behaviour & Attitude

Subcategory 2 – statistically confirmed
Referent: Female + Inanimate Concrete
Evaluation: Positive
ToD: fashion & decor, art & culture, health & appearance

Subcategory 3 – not confirmed
Referent: Female + Events- Activities & States of Affairs
Evaluation: Neutral
ToD: entertainment & fame, family & love, emotion & psychology, science & politics

In this study, the proof-of-principle that keyword analysis of abstract concepts can been performed in Multifactorial Usage-Feature Analysis has been demonstrated. This marriage of methodologies provides quantified corpus-driven results which can be falsified through repeat analysis and which are sensitive to social variation. In this study, social variation has been restricted to Topic of Discourse, but the extension of this to other extra-linguistic variables is methodologically straightforward. The main limitation in this regard is data sparseness. The manual analysis of the usage-feature analysis is laborious and time consuming, meaning that practical constraints limit the number of examples that can be annotated in a give study. We saw that small sample size dramatically restricts the possible complexity of the modelling. This is especially true for confirmatory analysis. However, the possibility for modelling conceptual-functional structures without a *tertium comparationis*, such a linguistic form or a 'conceptual metaphor' has been demonstrated. More complete studies that bring in different languages and / or langauge varieties or diachronic change, will still be needed to see if the increased complexity can be adequately modelled with the limited data available in manually annotated language samples.

## References

*Agresti, A.* (2013). *Categorical data analysis* (3rd ed.). New York: John Wiley.

*Bartmiński, J.* (2009). *Aspects of cognitive ethnolinguistics*. London: Equinox.

*Blum-Kulka, S., House, J. & Kasper, G.* (1989). *Cross-cultural pragmatics: Requests and apologies*. Norwood: Ablex.

*Davies, M.* (2008). *The corpus of contemporary American English: 450 million words, 1990-present.*

*Dirven, R., Goossens, L. Putsey, Y. & Vorlat, E.* (1982). *The scene of linguistic action and its perspectivization by speak, talk, say, and tell.* Amsterdam: John Benjamins.

*Divjak, D.* (2006). Ways of intending: A corpus-based cognitive linguistic approach to near-synonyms in Russian. In St. Th. Gries & A. Stefanowitsch (Eds.), *Corpora in Cognitive Linguistics. Corpus-based approaches to syntax and lexis* (pp. 19–56). Berlin: Mouton de Gruyter.

*Divjak, D.* (2010). *Structuring the lexicon: A clustered model for near-synonymy.* Berlin: Mouton de Gruyter.

*Divjak, D., & Gries, St. Th.* (2006). Ways of trying in Russian: Clustering behavioral profiles. *Corpus Linguistics and Linguistic Theory*, *2*, 23–60.

*Geeraerts, D., Grondelaers, S. & Bakema, P.* (1994). *Structure of Lexical Variation. Meaning, naming and context*. Berlin: Mouton de Gruyter.

*Glynn, D.* (2007). *Mapping meaning. Toward a usage-based Cognitive Semantics.* Leuven: University of Leuven.

*Glynn, D.* (2008). Lexical fields, grammatical constructions and synonymy. A study in usage-based Cognitive Semantics. In H.-J. Schmid & S. Handl (Eds.), *Cognitive foundations of linguistic usage-patterns. Empirical studies* (pp. 89–118). Mouton: Berlin: Mouton de Gruyter.

*Glynn, D.* (2009). Polysemy, syntax, and variation. A usage-based method for Cognitive Semantics. In V. Evans & S. Pourcel (Eds.), *New directions in Cognitive Linguistics* (pp. 77–104). Amsterdam: John Benjamins.

*Glynn, D.* (2010a). Testing the Hypothesis. Objectivity and verification in usage-based Cognitive Semantics. In D. Glynn & K. Fischer (Eds.), *Quantitative Cognitive Semantics. Corpus-driven approaches*. Berlin: Mouton de Gruyter, 239–270.

*Glynn, D.* (2010b) Corpus-driven Cognitive Semantics. Introduction to the field. In D. Glynn & K. Fischer (Eds.), *Quantitative Cognitive Semantics. Corpus-driven approaches* (pp. 1–42). Berlin: Mouton de Gruyter.

*Glynn, D.* (2014a). The many uses of *run*: Corpus methods and Socio-Cognitive Semantics. In D. Glynn & J. Robinson (Eds.), *Corpus methods for semantics. Quantitative studies in polysemy and synonymy* (pp. 117–144). Amsterdam: John Benjamins.

*Glynn, D.* (2014b). Correspondence Analysis. An exploratory technique for identifying usage patterns. In D. Glynn & J. Robinson (Eds.) *Corpus methods for semantics. Quantitative studies in polysemy and synonymy* (pp. 443–486). Amsterdam: John Benjamins.

*Glynn, D.* (2014c). Techniques and tools. Corpus methods and statistics for semantics. In D. Glynn & J. Robinson (Eds.), *Corpus methods for semantics. Quantitative studies in polysemy and synonymy* (pp. 307–342). Amsterdam: John Benjamins.

*Glynn, D.* (2014d). Polysemy and synonymy. Corpus method and cognitive theory. In D. Glynn & J. Robinson (Eds.), *Corpus methods for semantics. Quantitative studies in polysemy and synonymy* (pp. 7–38). Amsterdam: John Benjamins.

*Glynn, D.* (2014e). Conceptualisation of home in popular Anglo-American texts. A multifactorial diachronic analysis. J. Díaz (Ed.), *Diachronic Studies in Conceptual Metaphor* (265–294). Amsterdam: John Benjamins.

*Glynn, D.* (2014f) The social nature of anger: Multivariate corpus evidence for context effects upon conceptual structure. In I. Novakova, P. Blumenthal, & D. Siepmann (Eds.), *Emotions in discourse* (pp. 69–82). Frankfurt/Main: Peter Lang.

*Glynn, D. & Fischer, K. (Eds.).* (2010). *Quantitative methods in Cognitive Semantics: Corpus-driven approaches*. Berlin: Mouton de Gruyter.

*Glynn, D. & Robinson, J. (Eds.).* (2014). *Corpus methods for semantics. Quantitative studies in polysemy and synonymy.* Amsterdam: John Benjamins.

*Greenacre, M.* (2007). *Correspondence analysis in practice.* London: Chapman & Hall.

*Gries, St. Th.* (1999). Particle movement: A cognitive and functional approach. *Cognitive Linguistics, 10,* 105–145.

*Gries, St. Th.* (2003). *Multifactorial analysis in corpus linguistics: A study of particle placement.* Continuum: London.

*Gries, St. Th.* (2006). Corpus-based methods and cognitive semantics: the many meanings of *to run*. In St. Th. Gries & A. Stefanowitsch (Eds.), *Corpora in cognitive linguistics: corpus-based approaches to syntax and lexis* (pp. 57–99). Berlin: Mouton de Gruyter.

*Gries, St. Th.* (2010). Behavioral Profiles: A fine-grained and quantitative approach in corpus-based lexical semantics. *Mental Lexicon, 5,* 323–346.

*Gries, St. Th. & Divjak, D.* (2009). Behavioral profiles: A corpus-based approach towards cognitive semantic analysis. In V. Evans & S. Pourcel (Eds.), *New directions in Cognitive Linguistics* (57–75). Amsterdam: John Benjamins.

*Gries, St. Th., & Stefanowitsch, A. (Eds.).* (2006). *Corpora in Cognitive Linguistics: Corpus-based approaches to syntax and lexis.* Berlin & New York: Mouton de Gruyter.

*Janda, L. & Solovyev, V.* (2009). What constructional profiles reveal about synonymy: A case study of the Russian words for sadness and happiness. *Cognitive Linguistics, 20,* 367–393.

*Kövecses, Zoltán* (1986). *Metaphors of ANGER, PRIDE, and LOVE. A lexical approach to the structure of concepts.* Amsterdam: John Benjamins.

*Krawczak, K.* (2014a). Epistemic stance predicates in English: A quantitative corpus-driven study of subjectivity. In D. Glynn, & M. Sjölin (Eds.), *Subjectivity and epistemicity: Corpus, discourse, and literary approaches to stance* (pp. 355–386). Lund: Lund University Press.

*Krawczak, K.* (2014b). Shame and its near-synonyms in English: A multivariate corpus-driven approach to social emotions. In I. Novakova, P. Blumenthal & D. Siepmann (Eds.), *Emotions in discourse* (pp. 84–94). Frankfurt/Main: Peter Lang.

*Krawczak, K.* (2014c). Corpus evidence for the cross-cultural structure of social emotions: Shame, embarrassment, and guilt in English and Polish. *Poznań Studies in Contemporary Linguistics, 54,* 441–475.

*Krawczak, K. & Glynn, D.* (2011). Context and cognition. A corpus-driven approach to parenthetical uses of mental predicates. In K. Kosecki & J. Badio (Eds.), *Cognitive Processes in Language* (pp. 87–99). Frankfurt/Main: Peter Lang.

*Krawczak, K., & Kokorniak, I.* (2012). A corpus-driven quantitative approach to the construal of Polish think. *Poznań Studies in Contemporary Linguistics, 48*, 439–472.

*Lakoff, G.* (1987). *Women, fire, and dangerous things. What categories reveal about the mind.* Chicago: University of Chicago Press.

*Nordmark, H. & Glynn, D.* (2013). anxiety between mind and society. A corpus-driven cross-cultural study of conceptual metaphors. *Explorations in English Language and Linguistics, 1*, 107–130.

*Speelman, D. & Glynn, D.* (2005). *LiveJournal Corpus of American and British English.* University of Leuven.

*Sten, J. & Glynn, D.* (2011). The American concept of home. A multifactorial corpus-driven study. *Interstudia, 10*, 65–79.

*Vorkachev, S.* (2007). *Любовь как лингвокультурный концепт.* Moscow: Gnozis

*Wiebe, J., Wilson, T., & Cardie, C.* (2005). Annotating expressions of opinions and emotions in language. *Language Resources and Evaluation, 39*, 165–210.

*Wierzbicka, A.* (1985). *Lexicography and Conceptual Analysis.* Ann Arbor: Karoma.

*Wierzbicka, A.* (1991). *Cross-cultural pragmatics: The semantics of human interaction.* Berlin: Mouton de Gruyter.

Karolina Krawczak
*Faculty of English, Adam Mickiewicz University, Poznań, Poland*

# Negative self-evaluative emotions from a cross-cultural perspective: A case of 'shame' and 'guilt' in English and Polish

**Abstract:** This study investigates the socially grounded concept of SHAME in a cross-linguistic and cross-cultural context. The concept is operationalized through two adjectives instantiating it, namely *ashamed* and *guilty* in British English and American English and their respective equivalents in Polish. Since concepts such as SHAME are determined by a complex system of intersubjective assumptions and rules, it is expected that differences in their conceptualization will emerge across the three communities. Some of these divergences will have to do with the ideas of individualism and collectivism, as represented here by the Anglo-Saxon world and Poland, respectively. The approach adopted here combines detailed qualitative analysis of natural examples with multivariate quantitative modeling. This makes possible the identification of frequency-based patterns of language use, which, in turn, afford an insight into conceptual and socio-cultural models of the phenomenon under investigation.

Keywords: negative self-evaluative emotions, usage-based, manual qualitative annotation, quantitative statistics

## 1. Introduction

The present paper examines two specific instantiations of negative self-evaluative emotions, namely, the lexical categories of 'shame' and 'guilt', as realized by their adjectival exponents. It is a comparative cross-cultural and cross-linguistic study, whose primary goal is to investigate the conceptualization of these emotion categories in three distinct communities, i.e., British English, American English and Polish. These societies can be perceived as epitomizing two opposing world-views: the individualism of the Occidental, Protestant and capitalist world and the relative collectivism of the Eastern, Catholic, post-communist reality. These socio-cultural differences are, among others, reflected in the proneness to and structuring of social emotions. The exact nature of this link will be explored in what follows. The secondary goal taken up here is to further develop multivariate quantitative corpus methods for the study of social emotions (cf. Krawczak 2014a, 2014b).

In general terms, negative social emotions, subsumed under the generic concept of SHAME, are engendered by the subject's awareness of others' real or potential disapprobation of his/her actions and/or personal traits. Ultimately, this social awareness stems from the experiencer's enculturation in the community to which s/he belongs and in which s/he functions (Lewis 2008: 746). Given this, such negative self-conscious emotions as shame or guilt can be described as inextricably linked to and buttressed by a complex system of socio-cultural assumptions, expectations and conventions (Gilbert 2003: 1222; Barret 2005: 956; Lewis 2008: 743). They determine what is socially acceptable and what deserves condemnation. In this respect, social emotions such as shame or guilt can be taken to constitute a regulatory system, imposed on the subject (Kaufman 1996: xii; Edelstein & Shaver 2007: 205; Lewis 2008: 746).

This monitoring function that such emotions perform operates in a twofold manner. On the one hand, the check may be personal and come from within, making the subject feel an internal poignancy, whose emergence is largely independent of the outside world. The experiencer simply feels him-/herself that something s/he does or did or some properties that characterize him/her make him/her inferior socially or morally. There is no need for anyone to actually know of the misconduct or vice in question. On the other hand, such a self-check may be initiated interactively or socially, resulting from and depending on the immediate or potential presence of witnesses. In this latter case, the absence of an audience or the certainty that the shameful behavior or characteristic will never be revealed would mean that no negative self-evaluative emotion would arise. In the relevant literature, the former "mechanism" of internally-exerted "control" (Triandis et al. 1988: 326; cf. Edelstein and Shaver 2007: 202) tends to be identified as peculiar to guilt – understood as a relatively more "private" and "mature" emotion. By contrast, the latter process of externally-motivated regulation is taken to be more typical of shame, the more "public" of the two states (Tangney et al. 1996: 1256). Such a picture implies that more shame-prone subjects could be perceived as exhibiting less responsibility for oneself and less virtuousness, as it is only the actual or envisaged criticism of others that induces negative self-assessment.

This distinction, however, which appears to dichotomize the two emotions in a categorical manner, is not entirely accurate, as Kaufman (1996:6ff.) notes. This polarizing picture portrays shame as a more public, dependent and immature materialization of negative self-evaluative emotions, with guilt, on the other hand, surfacing as the more private, individuated and mature manifestation of such emotions. This juxtaposition further suggests that the two categories can be discreetly differentiated, which is not necessarily true. As a matter of fact, experimental studies (e.g., Tangney *et al.* 1996: 1257) have demonstrated that both shame and guilt can be felt as a result of identical situations, which shows that the two emotions are not necessarily categorically divergent with respect to their etiological context. It has been noted, nonetheless, that while "moral transgressions" were linked to both emotions, non-moral wrongdoings were more commonly, but by no means exclusively, associated with shame (Tangney *et al.* 1996: 1257).

The above observations are directly linked to another more general dualistic distinction, which is essential in the discussion of negative social emotions. It concerns cultural frameworks, habitual behavioral patterns and attitudinal profiles. More precisely, it relates to the ideas of collectivism and individualism (Hofstede 1980; Triandis 1995, 2001; Oyserman *et al.* 2002, 2008), which are claimed to run deep in the cultural discursive practices that motivate and structure the way individuals think about and interact with the surrounding world. In the present context, there are a number of characteristics of these two cultural frames that deserve our attention.

Collectivist communities are typically associated with an intrinsic interdependence and low social adjustability of its members, who attach a lot of weight to the stability of their relations and to their public image (Sznycer *et al.* 2012: 354). Given the inherent interrelatedness of such societies, this image could be described as extended, as it depends not only on the individual's own deeds and properties but also on those of his/her dear ones. Collectivistically-oriented people are led in their actions by what could be referred to as a societal guiding hand or an external signpost of right and wrong, materialized in the "judging eyes" of other people (Rochat 2009: 110). When there is no one to watch and this moral signpost is, thus, out of sight, collectivists are more likely to violate the externalized guidelines of dos-and-don'ts without any emotional repercussions. Collectivistic thinking is attributed to the eastern and southern world, which is here represented by the Polish society (e.g., Hofstede 1980; Triandis 1989; Oyserman & Lee 2008).[1] Given the postulated characteristics of such cultures, they are claimed to be more shame-oriented (Triandis *et al.* 1988: 326; Edelstein & Shaver 2007: 202), i.e., more likely to experience negative social emotions as a result of misdoings or defects that have been disclosed.

1 Incidentally, in the multi-national ratings on the scale of individualism–collectivism, Poland is situated in a relatively medial position, slightly closer to the latter end of the continuum, which means that it manifests features of both frameworks. However, among the three communities examined here, it undoubtedly is the most collectivistic.

Individualism, by contrast, is customarily linked to independence and high social flexibility (Sznycer *et al.* 2012: 354). What counts to individualistically-oriented people is not so much the judgment of others as their own internalized system of norms (Triandis 1989: 510; Markus & Kitayama 1991: 226; Triandis 2001: 909; Oyserman *et al.* 2002: 4). They are, therefore, liable to experience negative self-evaluation and the resultant emotions pertaining to the field of shame regardless of whether there should be anyone to witness their transgressions. This is because the system of standards they adhere to is not only socially motivated but also internalized and so they will feel bad whenever they infringe on their principles. In this respect, individualists are construed as guilt-oriented insofar as their experience of negative self-evaluative emotions is initiated from within (Triandis et al. 1988: 326; Edelstein & Shaver 2007: 202). This internally imposed regulation and its precedence over external restraints is also expressed in higher interpersonal flexibility, which means that their social anchors are weaker than those of collectivists. As a consequence, rather than withstanding social inconveniences that jeopardize their personal well-being, individualists are more likely to leave and move on (Sznycer *et al.* 2012: 354f.). This independent frame of mind is stereotypically ascribed to western cultures, especially the Anglo-Saxon world, with the United States of America figuring as the most individuated society (e.g., Oyserman et al. 2002; Triandis 1989).

On the basis of such observations, we might hypothesize that negative social emotions will be experienced in more collectivistic societies such as Poland as a result of revealed misdoings, in the presence of witnesses and will be directly linked to the interactive situations. It is also likely that other people's behavior or properties will be more likely to engender shame-emotions in collectivists. In more individualistic communities, on the other hand, such as those of Britain or America, social emotions may be expected to arise in relation to self's, rather than others', actions and regardless of whether the cause is publicly known or not.

## 2. Method and data

The method employed in the present study is termed "configurational" (Geeraerts *et al.* 1994, 1999), "profile-based" (Gries 2006; Gries & Stefanowitsch 2006) or "multifactorial usage-feature" (Glynn 2009, 2010) analysis.[2] The central premise of

2 A sizable collection of studies using this methodology can be found in Glynn and Fischer (2010) and Glynn and Robinson (2014). In addition, the latter volume provides a thorough tutorial for a number of quantitative methods that can be applied to categorical data.

this corpus-based methodology is that contextualized language structure provides access to conceptual organization, which, in turn, is an index of socio-cultural tendencies in construing the world. The goal in any such usage-based study is to find frequency-based patterns of language use that, in light of the above premise, will be indicative of the conceptual and cultural behavioral and attitudinal patterns. Accordingly, in order to identify such cognitive and cultural models of a given phenomenon under investigation, linguistic means of accessing it need to be established. Once such lexico-constructional types are selected, a substantial number of their contextualized instantiations are extracted from relevant corpora and analyzed qualitatively. This analysis consists in manual annotation for a wide spectrum of formal, semantic and sociolinguistic usage-features. The choice of such factors is determined by the specific research questions to be addressed and the hypotheses to be tested. These variables, informed by prior empirical and theoretical research, are instrumental in revealing the socio-cultural and conceptual frames underlying the object of study. These frames are identified through multivariate statistical modeling of the annotated data.

The present study further advances the work of Krawczak (2011, 2013, 2014a, 2014b), whose goal is to develop adequate corpus-based quantitative methodology for the description of abstract concepts such as SHAME, which are interpersonally emergent structures. It is also complementary to the GRID-based methodology as pursued in the experimental research gathered in Fontaine *et al.* (2013) or the study of collocational attraction of Lewandowska-Tomaszczyk and Wilson (2014).

The concept of SHAME is operationalized by the author through two adjectives realizing it, namely, *ashamed* and *guilty* in British and American English and *zawstydzony* and *winny* in Polish. Equal numbers of observations for each lexeme for each of the three linguistic communities were extracted from the fiction components of British National Corpus (BNC), the Corpus of Contemporary American English (COCA) and the National Corpus of Polish (NKJP). The data altogether amount to 450 observations, each accompanied by a broad context. All the examples were manually annotated for a set of semantic-functional features, as discussed in section 3.

## 3. Analysis

The contextualized examples were submitted to detailed manual analysis for a number of explanatory factors pertaining to the event structure of negative social emotions, as detailed in Table 1. The selection of these variables was determined by the author's prior studies (e.g., 2011, 2012, 2014a, 2014b) and the findings of research in cognitive or social psychology as well as linguistics (e.g., Kövecses

1986, 1990; Wierzbicka 1992; Triandis 1995; Tissari 2006; Fabiszak & Hebda 2007; Tangney & Tracy 2012).

*Table 1: Explanatory variables*

| *Explanatory factor* | *Usage-features* |
|---|---|
| *Audience* | *Present, Absent* |
| *Cause* | *(1) Bodily; (2) Dubious Status; (3) Failure: (a) to Others, (b) to Self; (4) Inadequacy; (5) Insecurity; (6) Norm Violation: (a) Decency, (b) Politeness, (c) Emotional Reaction; (7) Status Loss: (a) Financial, (b) Mistreatment; (c) Unprestigious Status.* |
| *Cause Time* | *General, Past, Present* |
| *Emotion Type* | *Individual, Shared* |
| *Status* | *Moral, Social* |

The most vital of these variables concerns the specific stimulus that causes the emotional experience to arise in the first place. There are eight main levels recognized here, three of which are further subdivided into more specific causes (see Table 1). Let us exemplify each of the causing factors with an adequate example.

(1) *This is exactly what his sister was telling him. She had been telling him this for years. She had never approved of Fredericka's obesity. She was* ***ashamed of*** *how it made Lorne appear, of what people were saying about a man who chose a woman who so thoroughly outsized him* (Bodily)

(2) *Then she felt desperately* ***guilty****. Who was she, who'd been crucified by Hamish's departure, to hanker after someone else's husband?* (Dubious Status)

(3) *She was to blame. The shock of finding out about her child had been responsible for her husband's fatal apoplexy, and she would feel* ***guilty*** *for the rest of her life.* (Failure to Others)

(4) *Now he's pulled me back in. He knew I wanted this boat and he used it and he got me working for him, which I swore I would never do. I feel* ***ashamed*** *because I had a price. He named it. And now I know that about myself.* (Failure to Self)

(5) *The worry which most perturbed Winnie was one of which she was deeply* ***ashamed****. She had found, since her return to the house, that she was horribly nervous of being alone in it at night.* (Inadequacy)

(6) *Ledwie dotknęli dzwonka, a drzwi na pierwszym piętrze otworzyły się. – Przyjechali! – radośnie wykrzyknęła na ich widok mała dziewczynka w czerwonej sukience z wielkim białym kołnierzem i* ***zawstydzona*** *zniknęła w zakamarkach mieszkania.* (Insecurity)

[No sooner did they touch the bell than the door on the first floor opened – They are here! – shouted happily a little girl wearing a red dress with a big white collar and ashamed she disappeared into the flat.]

(7) *" Oh sir, I am so **ashamed**. Why would they force me to do this so suddenly?" Her voice still quaked as she spoke. Dickens reached out his hands toward her shoulders to calm her, but quickly drew them back as if thinking better of the idea. " (...)" Who is forcing you to do what? " Dickens asked quietly. " " Only tonight, after the play had begun, and my mother was already onstage, he brought me this new costume. He ordered me to play the whole scene with both hands at my sides, as if I was too dumbstruck by the appearance of Lady Macbeth to notice my own indecency. " " Indecency? " Dickens said it with a slight catch in his voice. My eyes followed Dickens's eyes as they tried to solve the mystery of her costume, which she was holding bunched at the neck with that hand not occupied with Dickens's handkerchief. With the silent eloquence that only a natural actress could accomplish, she let both of her hands fall to her sides. The coarse brown peasant's smock fell open. Its neck hole had been slashed downward and the front was almost completely undone to her waist. She stood there helplessly, tears brimming in her eyes, her white shoulders and the tops of her breasts almost fully revealed. "Yes, I see, " Dickens delicately averted his eyes which, of course, caused me to avert my own.* (Norm Violation: Decency)

(8) *Pat also went to see Gildas and Ludens, both of whom were feeling **guilty** because they had not offered to' put him up'.* (Norm Violation: Politeness)

(9) *Nadspodziewanie Palmiak wybuchł spazmatycznym płaczem. Przyłożył pięści do oczu, rozmazywał łzy po fioletowych policzkach, usta miał otwarte, spróchniałe zęby tonęły w ślinie. Kasjerka, jakby **zawstydzona**, pochyliła się nad zeszytem, panny bufetowe i dziewczęta cofnęły się za drzwi.* (Norm Violation: Emotional Reaction)

[Unexpectedly, Palmiak burst out into spasmodic tears. He put his fists on his eyes, smearing the tears on his violet cheeks, with his mouth open, his decayed teeth drowning in saliva. The cashier, as if ashamed, bent over her notebook, the buffet attendants vanished behind the door].

(10) *Did she expect Adam to be **ashamed** of his lack of furnishings?* (Status loss: Financial)

(11) *she was drunk and hit me in front of that hateful boy Peter Catesby. I can remember standing there with her slap on my cheek and feeling **ashamed**, outraged, shocked, everything... but sorry for her.* (Status loss: Mistreatment)

(12) *I dread walking near them, **ashamed** of my loneliness.* (Status loss: Unprestigious Status)

Bodily causes relate to the physical appearance, outfit, physiological functions or diseases, as illustrated in (1). Dubious status (2) occurs when the subject does something that is wrong and that is a reason for concern irrespective of whether anyone other than the wrongdoer should ever find out about it. The next stimulus that may engender shame emotions has to do with failure to meet one's own standards and expectations (3) or those of others' (4). Inadequacies refer to certain mental, physical or interactive defects, which tend to be temporary lapses (5). Another socially-motivated cause has to do with insecurities, which cause momentary self-consciousness or discomfort, but which can be linked to either a fleeting loss of confidence or more deep-rooted anxieties (6). Yet another group of circumstances that may be the basis for negative self-evaluative emotions concerns violation of norms that regulate social conduct in terms of decency, politeness and appropriateness of emotional reaction ((7)–(9)). Finally, negative social emotions may be initiated by actual or perceived loss of status due to one's financial situation, mistreatment received at the hands of other people or being part of some social or professional group that lacks status ((10)–(12)).

The next set of usage-features for which the data were annotated concerns the temporal frame of the cause. Three such frames are identified here: general or atemporal causes (13), stimuli that appertain to the past (14), and those that are embedded in the present interactive situation (15):

(13) *He never seemed able to understand why God had made Minnie and me so small, and I believe he was slightly* ***ashamed*** *of us. Whenever we were out together as a family, he always kept his head bent; this way, he did not have to look anyone in the eye. I'm not sure he completely understood why he did this, or what he was afraid to encounter in the gaze of his fellow man; perhaps he simply didn't want to see pity for us there – or for himself.* (Cause time: General)

(14) *Mallory sucked in her belly and tried not to feel* ***guilty*** *about the cinnamon roll she'd inhaled on a quick break two hours ago.* (Cause time: Past)

(15) *Camille thought about the fact that the violence that had taken place in her home would eventually be played out publicly and her neighbors would soon know what happened. She cringed at the thought of being the only black family on the block and the first with a murder scene at their home. Camille felt* ***guilty*** *for thinking of her own embarrassment at a time like this. Misa had murdered her brother-in-law. Surely, her sister would be arrested tonight.* (Cause time: Present)

There are three more dimensions that are deemed essential to the description of negative social emotions. Firstly, a bipartite distinction is introduced between social emotions experienced on an individual basis, on the one hand, and as shared events, on the other. In the former case, the experiencer him-/herself is the source and target of the negative evaluation (14), while in the latter instance, such uncomfortable self- or, rather, other-consciousness is due to another person's behavior or properties but is extended to affect the experiencer (13). The next variable concerns the presence or absence of audience, as evidenced in (11) and (15), respectively. The final factor pertaining to the structure of negative social emotions has to do with the status of the emotional experience, which can be either moral (16) – arising due to deeds that are in conflict with the unwritten moral code of good and evil – or social (8) – initiated by infringement on some social conventions.

(16) *Świadczą o tym na przykład wyznania ludzi katujących swoich bliskich. Pierwsze uderzenia są spowodowane złością, potem jednak bijący zaczyna się czuć* ***winny*** *i – zgodnie z paradoksalnym efektem poczucia winy – uderza znowu, ponieważ poczucie winy osłabia jego wiarę w możliwość poprawienia i skontrolowania swojego zachowania.* (Status: Moral)

[It is evidenced, for example, by the confessions of people who beat their loved ones. The first blows are caused by anger, but later the beater starts to feel guilty and – in accordance with the paradoxical sense of guilt – he beats again, because the sense of guilt weakens his belief in the possibility of remedying and controlling his behavior.]

In the former case, where negative self-evaluative emotions stem from morally questionable conduct, the emotion is likely to have long-lasting consequences. On the other hand, when the causing factor is an instance of breaching some tacit social protocol, the experience is much less grievous and is most likely to vanish without a trace along with the event engendering it.

## 4. Results

This section presents the results of the quantitative analysis to which the data were submitted following their manual annotation. Two methods were used here, one exploratory in the form of correspondence analysis, the other confirmatory in the form of logistic regression analysis. The former method identifies frequency-based associations of usage-features in multidimensional data. The other method, in turn, serves to verify the revealed usage correlations by calculating their statistical significance, effect size and the overall predictive accuracy of the model.

### 4.1 Exploratory results: Identifying usage profiles

Let us first consider the exploratory results presented in Figure 1. This multiple correspondence analysis plot visualizes the associations between the following set of usage-features: the specific emotion category relative to its linguistic context, emotion cause, temporal frame of the cause, emotion type, status and audience. In this "space reduction" method, the proximity between data-points is indicative of the relative degree of their association (Glynn 2014: 443).

*Figure 1. Results of multiple correspondence analysis: Associations between Concept-Language and Audience, Cause, Cause Time, Emotion Type and Status.*

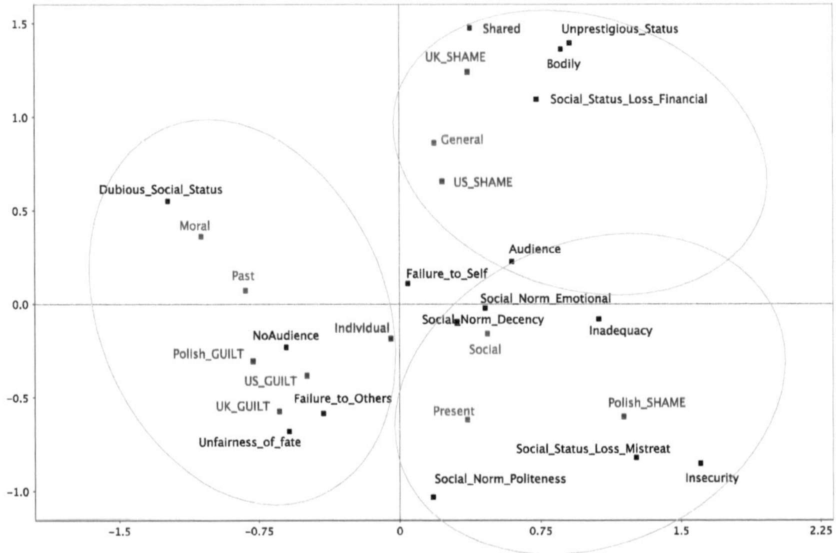


There are three distinct clusters emerging in the plot: one cluster for the lexical category of 'guilt', which is consistent across the three communities, and two groupings for 'shame', one representing the usage correspondences in the more independent communities of Britain and America, the other showing the usage profile for the relatively more collectivistic society of Poland. Let us now focus on the specific associations for each of the emergent clusters, starting with 'guilt'. Notwithstanding the linguistic context, the adjectival exponent of this lexical category is related in usage to three sets of causes, each comparatively severe in its impact upon the experiencer and possibly his/her social environment. Two of these causes, i.e., unfairness and failing others, both of which are close to the data-points designating the actual lexemes, constitute distinct correlations. The third cause, namely, dubious status, which is located in another quadrant of the plot, can be assumed to be simultaneously attracted to the usage profile of Anglo-Saxon 'shame'. It can, therefore, be regarded as a more distant correspondence here. The other associations surfacing in this grouping relate 'guilt' to the absence of audience and the individual type of the emotion. Both these features are consonant with the claim that this emotion is relatively more private of the two and so is more likely to be experienced in solitude and as a result of one's own doings. The adjectives instantiating 'guilt' in the three communities are also linked to two more features, i.e., actions originating in the past that are morally, rather than socially, questionable. Overall, the picture that emerges here for this lexical category is congruent with the assumption that it is a much heavier emotional experience (see section 1).

In the right-hand half of the bi-plot there are the two other clusters, one for the Anglo-Saxon usage profile of *ashamed*, the other for the behavioral tendencies of its Polish equivalent, *zawstydzony*. Let us begin with the former. In both British and American English, the lexical category 'shame' is closely associated with three types of stimuli causing the emotion, i.e., bodily conditions, financial problems or being stigmatized as inferior <Unprestigious Status>. All these causes can be regarded as long-lasting states, which is further supported by their co-occurrence with the usage-feature of a general temporal scope of the cause. There are two more usage characteristics pertaining to this cluster and defining the usage profile of Anglo-Saxon 'shame', both located on its peripheries. The first association, the shared type of the emotion, obtains more typically for the British occurrences of *ashamed*, being located right above this data-point. The other feature, in turn, i.e., the presence of audience, is shared between the American instantiation of 'shame' and the Polish 'shame' cluster, being situated in between the two. This shows that relative to 'guilt', 'shame' especially in American English and Polish, but also, to a lesser degree, in British English, is experienced in the presence of witnesses. This is in line with those psychological findings that construe shame as a more public emotion (see section 1). Let us now see the other correspondences holding for *zawstydzony*.

In Polish, 'shame' is distinctly linked to causes that are clearly based in the immediate interactive situation, having to do with mistreatment, insecurities and violating the social norms of politeness. Less distinct relations obtain between *zawstydzony* and three other causes, i.e., indecent behavior, inappropriate emotional reaction or inadequacies. All of them, however, are clearly social in nature and require witnesses, which is substantiated here by the relevant usage-features of the social status of the emotion, the present temporal frame of the event and the already mentioned presence of audience. The usage-feature of failing oneself, which is located close to the middle of the plot, is not distinctly associated with any of the clusters, which indicates that it is equally shared between all the groupings.

Overall, the exploratory findings reveal some usage characteristics that are peculiar to the lexical categories irrespective of the language and other associations that are language-specific. With regard to the former results, we observe that regardless of the linguistic and cultural context, the lexical category of 'shame' is used to denote emotional states that are experienced communally, i.e., in the presence of other people, and in relation to the violation of social norms. That of 'guilt', by contrast, emerges as a more intimate self-evaluative emotion, being experienced solitarily and having moral undertones. This is consistent with the profiles of the two emotions proposed in the relevant psychological literature, as discussed in section 1. With respect to the results that are culture-specific, there are some differences that emerge for the usage profiles of the adjectives realizing 'shame'. On the one hand, in the independent societies of Britain and America, 'shame' is experienced as a result of states that are not definable in terms of time, being general in nature, such as permanent physical conditions. In the relatively more interdependent community of Poland, by contrast, 'shame' is clearly linked to the present interactive situation where the emotion is initiated. Both these results correspond to the cultural assumptions put forward in section 1. One of the findings, nonetheless, clearly diverges from the expected profiles. It concerns the distinct association between British *ashamed* and the shared type of the emotion, a feature which was assumed to be more typical of collectivistic contexts.

## 4.2 Confirmatory results: Verifying usage profiles

Let us now consider the results of the confirmatory analysis conducted through multinomial logistic regression modeling. Six models were run, one for each level of the explained variable. Table 2 presents the statistically significant positive and negative correlations for the lexical categories under investigation in their respective linguistic contexts. The first column enumerates the independent variables (predictors) that emerged as statistically significant in at least one of the six models. The subsequent six columns specify the statistical significance and effect size for the positive and negative correlations between the relevant level of the independent variables and the given level of the dependent variable relative to a particular reference level specified in the top row.[3]

3 A reference level is a yardstick against which the behavior of the other elements in the sample is compared. It is important to note here that the positive and negative correlations for the level of the explained variable that serves as the reference level in a given model are not visualized in the output.

*Table 2: Results of multiple logistic regression modeling (package: Mlogit)*
*Language Concept ~ Cause + Cause Time + Emotion Type + Status + Audience*

**Models arranged by reference level (R.L)**

| | Model 1<br>R.L. – PL GUILT | Model 2<br>R.L. – PL SHAME | Model 3<br>R.L. – UK GUILT | Model 4<br>R.L. – UK SHAME | Model 5<br>R.L. – US GUILT | Model 6<br>R.L. – US SHAME |
|---|---|---|---|---|---|---|
| **Significant predictors with effect sizes for the given explained variable** | | | | | | |
| Shared emotion | US GLT: -1.83 . | US GLT: -1.87 . | ——— | US GLT: -2.48 *<br>US SHM: -7.98 . | PL SHM: 1.87 .<br>UK SHM: 2.49 * | UK SHM: 7.98 . |
| Past causes | US GLT: -9.02 . | ——— | ——— | ——— | PL GLT: 8.61 . | ——— |
| Present causes | UK SHM: -1.11 * | UK SHM: -1.88 ***<br>US SHM: -1.64 *** | UK SHM: -1.74 ***<br>US SHM: -1.36 ** | PL GLT: 1.08 *<br>PL SHM: 1.88 ***<br>UK GLT: 1.64 **<br>US GLT: 1.62 *** | UK SHM: -1.62 ***<br>US SHM: -1.38 ** | PL SHM: 1.63 ***<br>UK GLT: 1.40 **<br>US GLT: 1.38 ** |
| Social status | UK GLT: -1.32 * | ——— | PL GLT: 1.30 . | ——— | ——— | ——— |
| Audience | PL SHM: 2.23 ***<br>UK SHM: 1.82 ***<br>US GLT: -7.44 .<br>US SHM: 1.55 *** | PL GLT: -2.19 ***<br>UK GLT: -2.79 ***<br>US GLT: -3.00 *** | PL SHM: 2.74 ***<br>UK SHM: 2.44 ***<br>US SHM: 2.05 *** | PL GLT: -1.76 ***<br>UK GLT: -2.35 ***<br>US GLT: -2.57 *** | PL GLT: 8.04 .<br>PL SHM: 3.00 ***<br>UK SHM: 2.57 ***<br>US SHM: 2.30 *** | PL GLT: -1.50 ***<br>UK GLT: -2.09 ***<br>US GLT: -2.30 *** |
| Cause: Dubious Status | PL SHM: -3.24 *<br>US SHM: -2.52 . | PL GLT: 3.11 *<br>UK GLT: 3.11 *<br>UK SHM: 1.69 .<br>US GLT: 2.07 * | ——— | PL SHM: -1.69 . | PL SHM: -2.07 * | ——— |
| Cause: Failure to Others | PL SHM: -3.47 **<br>UK SHM: -3.39 **<br>US GLT: -1.99 .<br>US SHM: -4.52 *** | PL GLT: 3.41 **<br>UK GLT: 2.33 . | US SHM: -3.13 * | PL GLT: 3.29 **<br>UK GLT: 2.21 . | US SHM: -2.53 * | PL GLT: 4.42 ***<br>UK GLT: 3.34 *<br>US GLT: 2.54 * |
| Cause: Politeness | ——— | UK GLT: 2.87 * | PL SHM: -2.87 *<br>UK SHM: -3.29 *<br>US SHM: -2.45 * | UK GLT: 3.29 * | ——— | UK GLT: 2.46 * |
| Cause: Unfairness | US GLT: -2.58 . | ——— | US GLT: -2.79 * | ——— | PL GLT: 2.55 .<br>UK GLT: 2.72 * | ——— |
| **Model Stats** | C = 0.84 | C = 0.86 | C = 0.82 | C = 0.78 | C = 0.80 | C = 0.75 |

Signif. codes: 0 ‚***' 0.001 ‚**' 0.01 ‚*' 0.05 ‚.' 0.1 ‚ ' 1
Category/Language codes: PL GLT, PL SHM, UK GLT, UK SHM, US GLT, US SHM
———: lack of significant correlations

Before looking at the specific correlations, two general observations should be made. Firstly, there are three usage-features that emerge as significant irrespective of the reference level, i.e., the present frame of the cause, audience and failure to others as a specific stimulus giving rise to negative social emotions. This finding may be taken to indicate that these three dimensions play a central role in the intra- and inter-categorial structuring of both 'shame' and 'guilt', being thus particularly crucial for their description. It is noteworthy that both <Present Causes> and <Audience> relate directly to the hypotheses formulated in section 1. Another general point that deserves our notice is that the performance of the second model, where the reference level is 'shame' in Polish, is slightly better with C at 0.86. It is, nonetheless, a rather marginal difference. Overall, all the six models perform exceptionally well with the C statistic above 0.70. Let us now turn to the specific results for each of the lexical categories, starting with *winny* ('guilty') in Polish.

Not surprisingly, among the most important predictors for the usage profile of this lexeme are <Audience> and <Failure to Others>. In the former case, *winny* ('guilty') is disassociated from the presence of witnesses when its behavior is juxtaposed with that of 'shame' in Polish, British English and American English. When considered relative to 'guilt' in American English, on the other hand, it is positively correlated with this usage feature, with the effect size of this attraction being as high as 8.04. These findings reveal a clear consistency. On the one hand, *winny* ('guilty') stands out as the more individuated experience, unlike 'shame' in any of the three communities. On the other hand, when contrasted with the American instantiation of 'guilt', the Polish equivalent acquires a strong interdependent character, which is in line with the distinction introduced in section 1 between a relatively more collectivistic Poland and highly individualistic America. This is further supported by the clear disassociation of US *guilty* from this usage feature in as many as four of the models, relative to *winny* ('guilty') as well as 'shame' across the three communities. In a similar vein, *guilty* in British English is also disassociated from the presence of witnesses relative to the profile of 'shame' in the three linguistic groups.

The other usage characteristic, i.e., <Failure to Others>, is a distinct positive correlation for *winny* ('guilty') when its behavioral tendencies are compared to those of the Polish, British and American exponents of 'shame'. The same positive associations are revealed for *guilty* in British English. For *guilty* in American English, by contrast, a parallel link is found only when it is juxtaposed with American *ashamed*. Interestingly, 'shame' in American English is consistently disassociated from this usage-feature when compared to 'guilt' in any of the three communities, while in Polish or British English this only holds true for 'shame' when it is considered relative to *winny* ('guilty'). Overall, these correlations show that when the reference point is 'shame', this usage-feature is more readily associated with 'guilt', which might be related to the higher degree of maturity ascribed to this emotion (see section 1).

In addition to the above associations, 'guilt' in Polish is also strongly related to causes originating in the past when the reference level is *guilty* in American English. This latter exponent of 'guilt', in turn, is notably repulsed by this usage-feature when considered against its Polish equivalent. Furthermore, *winny* ('guilty') is linked to dubious status and unfairness as the causing conditions, characteristics that emerge when the reference level of comparison is *zawstydzony* ('ashamed') and US *guilty*, respectively. Thus, relative to 'guilt' in American English, 'guilt' in Polish is conceptualized in a more socially-involved manner, where injustice experienced by others is a cause for concern. On the other hand, when juxtaposed with the conceptualization of 'shame' in Polish, that of 'guilt' is clearly linked to moral transgressions <Dubious Status> that cast a shadow over one's social status. Weaker positive correlations obtain between *winny* ('guilty') and causes embedded in the present interactive situation and the social status of the emotion, with the reference levels being 'shame' and 'guilt' in British English, respectively. The situation is reversed when the Polish instantiation of 'guilt' is the reference level, in which case UK *ashamed* is disassociated from the present scope of the cause, while UK *guilty* emerges as clearly moral, rather than social, in its status. This shows that relative to the British exponents of 'guilt' and 'shame', *winny* ('guilty') is used in contexts that are less moral in their undertones and more interactive. This, in turn, may be taken to indicate that relative to the British society, in the Polish community, the category of 'guilt' is conceptualized in a manner that highlights the interdependence of the experiencer with respect to other people, who are likely to both witness his/her behavior and suffer its consequences.

Turning to 'guilt' in American English, we can see that in addition to the characteristics already mentioned above, it is also clearly disassociated from the shared type of the emotion when the reference levels are 'guilt' and 'shame' in Polish and 'shame' in British English. This again points to the comparatively more independent or individuated character of the conceptualization of this category in the American community, adding further quantitative evidence to the relatively higher degree of individualism of this group, particularly when compared to Poland. When juxtaposed with 'guilt' in either Polish or British English, US *guilty* is also clearly disassociated in usage from causes related to perceived unfairness of fate, where the self is advantaged and others are disadvantaged. This might suggest a more self-centered attitude characterizing Americans, which would be in line with the findings of experimental studies in which the highest level of individualism is ascribed to the American society. Finally, when contrasted with the Polish construal of 'shame' in Model 2, *guilty* in American English is positively correlated with dubious status as a cause of the emotion. This is a feature shared in Model 2 also with the exponents of 'guilt' in British English and Polish and with British *ashamed*, which tells us, in fact, more about the behavioral tendencies of *zawstydzony* ('ashamed') in Polish.

In British English, apart from the usage-features discussed above, *guilty* is also linked to breaching the social norms of politeness, when its behavior is compared to that of 'shame' across the three communities. Similarly to *winny* in Polish, but as opposed to *guilty* in American English, the UK exponent of 'guilt' is related in use to unfairness as a causing factor, when the reference level is that of US *guilty*. This finding brings forth an interesting dimension distinguishing these two Anglo-Saxon communities, where the society of Britain emerges as more socially or intersubjectively engaged and less individualistic. Further attention is required, nonetheless, to explain why *guilty* in British English, especially when contrasted with 'shame', should be linked to issues of politeness.

Finally, turning our attention to 'shame', on top of the usage-characteristics that have already emerged in our discussion of the profiling of 'guilt', there are a few more observations to be made. Firstly, an important dimension that clearly differentiates the Anglo-Saxon conceptualization of the category from its Polish model concerns the present frame of the cause. This feature emerges as a positive association for Polish, when *zawstydzony* ('ashamed') is juxtaposed with *ashamed* in British and American English, while being a negative correlation for the occurrences of the adjective in both UK and US, particularly when considered relative to 'guilt' across the three linguistic contexts. This clearly shows that the profile of Polish 'shame', unlike that of its Anglo-Saxon equivalent, is more interactive, being embedded in the here-and-now. Another essential dimension of comparison is the usage-feature <Audience>, which surfaces as a positive correlation for the three instantiations of 'shame' when the point of comparison is 'guilt'. This finding supports the assumption that the emotion of guilt is more likely to be experienced in seclusion (see section 1). A further commonality concerns the violation of social norms regulating politeness, which, irrespective of the linguistic and cultural context, is disassociated from 'shame', when the category is juxtaposed with the usage-profile of *guilty* in British English. This systematic disassociation is particularly informative here of the behavioral tendencies of the adjectival exponent of UK 'guilt', which, unlike any of the instantiations of 'shame', is positively correlated with breaching politeness norms. This is an interesting correlation, which should be further investigated. The other usage-patterns emerging here for 'shame' have to do with two other characteristics, i.e., <Shared Emotion> and <Dubious Status>. Let us discuss them in turn.

With regard to the shared type of the emotion, an interesting parallelism is revealed. The exponents of 'shame' in British English and Polish are both associated with this usage-feature. In the former case, this correspondence obtains when UK *ashamed* is compared to the behavioral tendencies of American *guilty* and *ashamed*, while in the latter instance, this only holds true when *zawstydzony* ('ashamed') is juxtaposed with *guilty* in American English. What this shows is that relative to 'guilt' and, to a lesser degree, 'shame' in American English, 'shame' in the two other communities is construed in a more collectivistic manner, being experienceable as a result of other people's actions. This comes as no surprise in the case of Polish, which has been expected to figure relatively high on the scale of collectivism, but it does call for further analysis with regard to the British community, where, on account of its presumed individualism, such a correlation was not anticipated. In American English, in accordance with what we would expect, *ashamed*, considered in comparison to its British equivalent, is disassociated from this usage-characteristic.

Finally, as regards <Dubious Status>, it emerges as a negative correlation for the profiles of 'shame' in Polish and American English and a positive correlation for British English. In Polish, the two are disassociated relative to 'guilt' in Polish and American English and 'shame' in British English. The American occurrences of *ashamed*, in turn, are negatively correlated with <Dubious Status> when the reference level is *winny* ('guilty'), which is thus a common denominator of this disassociation for the two communities. In British English, on the other hand, <Dubious Status> is associated with *ashamed* when the point of comparison is *zawstydzony* ('ashamed'), thus rendering the UK profile of 'shame' the most morally-bound of the three.

Overall, the usage-tendencies emerging for 'guilt' and 'shame' are consistent with the hypotheses and assumptions put forward in section 1. The lexical category of 'guilt', largely irrespective of the linguistic context, is conceptualized as a more individual emotion experienced privately, while 'shame' emerges as a more interactive experience associated with the presence of witnesses. Some of the results, however, reveal unexpected patterns of language use. Among these is the clear correlation of 'shame' in British English with the shared type of the emotion, which confirms the exploratory findings yielded in Figure 1, but which is at odds with the independence ascribed to this society within the theoretical framework of individualism. Similarly, the association of 'guilt' in Polish with social status and audience, whereby this category acquires such unforeseen interactive characteristics, diverges from the predicted picture. However, an interesting conditioning factor in both these cases is that the point of comparison is a community that is regarded as more individuated than the one being compared. Thus, the Polish conceptualization of 'guilt' emerges as clearly more collectivistic when juxtaposed with the same category in the Anglo-Saxon societies, while the British usage model of *ashamed* manifests more interdependent features when contrasted with the behavioral tendencies of negative social emotions in American English. This correlation renders these seemingly uninterpretable and counter-intuitive results clear, while also demonstrating the importance of the theoretical framework of individualism and collectivism in the description of the conceptual structure of self-evaluative emotions.

## 5. Conclusions

The present study has demonstrated that abstract concepts grounded in intersubjective experience such as SHAME can be described in a systematic manner with the use of corpus-based multivariate modeling. The picture thus obtained is informative not only with respect to the conceptual-lexical structuring of shame, but also with regard to the cross-cultural differences in profiling this concept. Moreover, the frequency-based usage tendencies revealed for the lexical categories relative to the linguistic and cultural contexts of their use and the specific register are statistically significant and verifiable, thus allowing us to make statements that generalize beyond the analyzed sample. The generalizations that emerged across the many usage-events support the hypothesized models ascribed to 'shame' and 'guilt' on the basis of the relevant literature in cognitive and social psychology, while also corroborating the importance of the theoretical framework of individualism and collectivism to the understanding of negative social emotions.

## References

*Barret, K. C.* (2005). The origins of social emotions and self-regulation in toddlerhood: New evidence. *Cognition and Emotion* 19(7), 953–979.

*BNC. Davies, Mark. (*2004–2013). BYU-BNC. (Based on the British National Corpusfrom Oxford University Press). < http://corpus.byu.edu/bnc/>

*COCA. Davies, Mark. (*2008–2013). The Corpus of Contemporary American English: 450 million words, 1990-present. http://corpus.byu.edu/coca/

*Edelstein, R.S. & P.R. Shaver.* (2007). A cross-cultural examination of lexical studies of self-conscious emotions. In Tracy, J.L., R.W. Robins & J.P. Tangney (Eds.), *The self-conscious emotions: Theory and research* (pp. 194–208). New York: Guilford.

*Fabiszak, M. & A. Hebda.* (2007). Emotions of control in Old English: Shame and guilt. *Poetica* 66, 1–36.

*Fontaine, J.R., K.R. Scherer & C. Soriano.* (2013). *Components of emotional meaning*. Oxford: OUP.

*Geeraerts, D., S. Grondelaers & P. Bakema.* (1994). *The structure of lexical variation. Meaning, naming and context*. Berlin: Mouton.

*Geeraerts, D., S. Grondelaers & D. Speelman.* (1999). *Convergentie en divergentie inde Nederlandse woordenschat. Een onderzoek naar kleding- en voetbaltermen.* Amsterdam: Meertensinstituut.

*Gilbert, P.* (2003). Evolution, social roles, and the differences in shame and guilt. *Social Research* 70(4), 1205–1230.

*Glynn, D.* (2009). Polysemy, syntax and variation. A usage-based method for Cognitive Semantics. In Evans, V. & S. Pourcel (Eds.), *New directions in Cognitive Linguistics* (pp. 77–106). Amsterdam: Benjamins.

*Glynn, D.* (2010). Synonymy, lexical fields, and grammatical constructions. A study in usage-based Cognitive Semantics. In Schmid, H.-J. & S. Handl

(Eds.), *Cognitive foundations of linguistic usage-patterns* (pp. 89–118). Berlin: Mouton.

*Glynn, D.* (2014). Correspondence analysis. Exploring correlations in multifactorial data. In Glynn, D. & J. Robinson (Eds.), *Corpus methods for semantics: Quantitative studies in polysemy and synonymy* (pp. 443–486). Amsterdam: Benjamins.

*Glynn, D. & K. Fischer (Eds.).* (2010). *Quantitative methods in Cognitive Semantics: Corpus-driven approaches*. Berlin: Mouton.

*Glynn, D. & J. Robinson (Eds.).* (2014). *Corpus methods for semantics: Quantitative studies in polysemy and synonymy*. Amsterdam: Benjamins.

*Gries, S.T.* (2003). *Multifactorial analysis in corpus linguistics: A study of particle placement*. London: Continuum.

*Gries, S.T. & A. Stefanowitsch (Eds.).* (2006). *Corpora in Cognitive Linguistics: Corpus-based approaches to syntax and lexis*. Berlin: Mouton.

*Hofstede, G.* (1980). *Culture's consequences*. Beverly Hills, CA: Sage.

*Kaufman, G.* (1996). *The psychology of shame. Theory and treatment of shame-based syndromes*. New York: Springer.

*Kövecses. Z.* (1986). *Metaphors of anger, pride, and love. A lexical approach to the structure of concepts*. Amsterdam: Benjamins.

*Kövecses. Z.* (1990). *Emotion concepts*. New York: Springer.

*Krawczak, K.* (2011). A socio-cognitive, corpus-driven analysis of intersubjective emotions: A case of shame in English. Paper presented at the 4th International Conference of the French Cognitive Linguistics Association, Lyon.

*Krawczak, K.* (2013). Construing social emotions in English and Polish: A quantitative corpus-driven approach to shame. Paper presented at the 5th International Conference of the French Cognitive Linguistics Association, Lille.

*Krawczak, K. (*2014a). Shame and its near-synonyms in English: A multivariate corpus-based approach to social emotions. In Blumenthal, P., I. Novakova & D. Siepmann (Eds.), *Les emotions dans le discours* (pp. 83–94). Frankfurt am Main: Lang.

*Krawczak, K. (*2014b). Shame, embarrassment and guilt: Corpus evidence for the cross-cultural structure of social emotions. *Poznań Studies in Contemporary Linguistics* 50(4), 441–475.

*Lewandowska-Tomaszczyk, B. & P. Wilson.* (2014). Self-conscious emotions in collectivistic and individualistic cultures: A contrastive linguistic perspective. In Romero-Trillo, J. (Ed.), *Yearbook of corpus linguistics and pragmatics 2014: New empirical and theoretical paradigms* (pp. 123–148).

*Lewis, M.* (2008). Self-conscious emotions: Embarrassment, pride, shame and guilt. In Lewis, M., J.M. Haviland-Jones & L.F. Barrett (Eds.), *Handbook of emotions* (742–756). New York: Guilford.

*McFadden, D.* (1978). Quantitative methods for analyzing travel behavior of individuals: Some recent developments. In Hensher, D.A. & P.R. Stopher (Eds.), *Behavioral travel modeling* (pp. 279–318). London: Croom Helm.

*NKJP. P. Pęzik.* (2012). Wyszukiwarka PELCRA dla danych NKJP [PELCRA searchengine for the NKJP data]. In Przepiórkowski, A., M. Bańko, R. Górski & B. Lewandowska-Tomaszczyk (Eds.), *Narodowy Korpus Języka Polskiego* [The NationalCorpus of the Polish language] (pp. 253–273). Warszawa: PWN.

*Oyserman, D., H.M. Coon & M. Kemmelmeier.* (2002). Rethinking individualism and collectivism: Evaluation of theoretical assumptions and meta-analyses. *Psychological Bulletin* 128(1), 3–72.

*Oyserman, D. & S.W.S. Lee.* (2008). Does culture influence what and how we think? Effects of priming individualism and collectivism. *Psychological Bulletin* 134(2), 311–342.

*Rochat, P.* (2009). *Others in mind. Social origins of self-consciousness.* Cambridge: CUP.

*Sznycer, D., K. Takemura, A.W. Delton, K. Sato, T. Robertson, L. Cosmides & J. Tooby.* (2012). Cross-cultural differences and similarities in proneness to shame: An adaptationist and ecological approach. *Evolutionary Psychology* 10(2), 352–370.

*Tangney, J.P., R.S. Miller, L. Flicker & D. Hill Barlow.* (1996). Are shame, guilt, and embarrassment distinct emotions? *Journal of Personality and Social Psychology* 70(6), 1256–1269.

*Tangney, J.P. & J. Tracy.* (2012). Self-conscious emotions. In Leary, M. & J.P. Tangney (Eds.), *Handbook of self and identity* (pp. 446–480). New York: Guilford.

*Tissari, H.* (2006). Conceptualizing shame: Investigating uses of the English word shame, 1418–1991. In McConchie, R.W., H. Tissari, O. Timofeeva & T. Säily (Eds.), *Selected proceedings of the 2005 symposium on new approaches in English historical lexis* (pp. 143–154). Somerville, MA: Cascadilla Proceedings Project.

*Triandis, H.C.* (1995). *Individualism and collectivism.* Boulder: Westview.

*Triandis, H. C.* (2001). Individualism-collectivism and personality. *Journal of Personality* 69, 907–924.

*Wierzbicka, A.* (1992). Defining emotion concepts. *Cognitive Science* 16, 539–581.

Naoki Nakamata
*Kyoto University of Education, Japan*

# A Categorization of conditional Expressions in Japanese: insights from a lexical approach

**Abstract:** It is generally agreed in the recent literature that collocation plays an important role in language learning. The term collocation can be expanded to include the relationship between function words and verbs. This paper researched the top 20 verbs that occur before four conditional markers in Japanese: *to*, *ba*, *tara*, and *nara*. Three of these markers—*to*, *ba*, and *tara*—show similar results. Interestingly, the verbs can be divided into five groups according to meaning: (1) verbs of perception or information processing, such as "see" and "compare"; (2) verbs of saying and thinking, such as "say" or "think"; (3) verbs of time or change, such as "become" or "begin"; (4) verbs of movement, such as "go to" or "go in"; and (5) verbs of existence, such as "be" or "have." This new categorization is helpful for learners to imagine using markers in daily life.

Keywords: Collocation, Verb, Conditional Marker, Subcategory, Corpus-Driven Approach

## 1. Introduction

What is the state of "being able to use an expression" for language learners? Simply knowing the meaning or understanding the grammatical rules of an expression is not sufficient. Woodland (2000:31) claims that "learning more vocabulary is not just learning new words, it is often learning familiar words in new combinations." Such combinations are known as *collocations*.

The word collocation brings to mind verb-noun combinations, such as *have a break* or adjective-noun combinations, such as *a sound sleep*. However, the term collocation can be expanded to include combinations of functional words and verbs, particularly in agglutinative languages such as Japanese.

Frequency is another important factor in language learning. The important question is not only "which combination is possible" but also "which combination is used most" since occasionally more than one combination is possible. Moreover, as Bybee (2010) notes, frequency is considered a part of grammar. Grammar for learners should contain rich information regarding collocations and their frequency.

This paper argues that examining the vocabulary that collocates with function words, which is called the *Lexical Approach* in this paper, gives us new insights into language studies and teaching. It will be demonstrated that the lexical approach supplies not only an inventory of useful expressions, but also a brand new description or classification of the grammar. As an example, this presentation deals with conditional expressions in Japanese. Conditional expressions have been studied in depth, so it is difficult to discover anything new using the usual approach. However, researching verbs that collocate with conditional markers enables us to re-categorize conditional expressions.

## 2. Basic Description of Japanese Grammar

This section describes some of the basics of Japanese grammar. Japanese is an agglutinative language; it contrasts with isolating or inflectional languages in that, in Japanese, a verb can connect with many kinds of functional words to express an extended meaning.

For example, consider the verb *ugoku* 'move.' Below are two examples of its extended meaning.

(1) ugo-i=teiru 　'[it] is moving'
(2) ugok-a=nai 　'[it does] not move'

To express the progressive aspect, as in 'it is moving,' the verb *ugoku* combines with the functional word *te-ir-u*[1], to become *ugoiteiru*. The verb stem is *ugok-*, and Japanese verbs require the insertion of a vowel, in this case *-i-*, to combine with function words. Moreover, in the course of historical change, the final consonant *k* is deleted, thus forming *ugo-i-teiru*. To create a negative sentence, *na-i* is attached to *ugoku*, resulting in *ugok-a-nai*, which means 'it does not move.' The above are typical grammatical categories, i.e., aspect and negation, but other types also exist.

(3) ugok-u toki 　'when [it] moves'
(4) ugok-i-sugiru 　'[it] moves too much'

Example (3) is a case of a subordinate clause, while example (4) is a case of a complex verb. Thus, function words that combine with verbs play a dominant role in sentence formation in Japanese grammar.

1 *Teiru* originally consists of the particle *te* and existential verb *ir-u*, so its structure is displayed as *te-ir-u*. The less detailed structure is chosen for convenience.

## 3. Previous Studies

Similar to other collocations such as collocations of verbs and nouns, the occurrence of verbs that collocate with a target item is not uniform; a small number of verbs occur frequently while most verbs occur infrequently, in accordance with Zipf's law. The lexical approach counts the observed instances of collocations of substance words with the target function word.

This approach provides benefits for both language learning and linguistic study. In learning, as Lewis (2000) has claimed, learners need to know collocations in order to use the target language. The important point is which combination is used most frequently. Most learners and teachers must have had questions such as, "Which phrases do native speakers use most?" or, "Do native speakers really use this sentence from the textbook?" Surveys based on corpora can answer these questions.

As an example of the lexical approach, Nakamata (2011) surveyed verbs that occur with the resultative aspect marker *tearu*. Theoretically, *tearu* can combine with any transitive verb. Surprisingly, however, 50% of the instances of *tearu* that appear in the corpora combine with a single verb, *kaku* 'write.' Therefore, *kaitearu*, which means 'be written,' accounts for 50% of the occurrences of *tearu*. While this may simply be a number for linguists, it provides great insight for language teachers, particularly when they have a lesson on *tearu* the following day. However, Nakamata (2011) only surveyed collocations of one function word; further surveys are necessary to confirm whether fruitful information for language learning and education can be garnered from collocation with other function words.

Concentration on frequent collocations can shed new light on descriptive studies of grammar. Bybee (2010) examined collocations with *can* and *can't* and found, for example, that a certain type of verb appears more often in negative sentences. She concluded that "frequently repeated structure can become conventionalized, and therefore part of grammar" (Bybee 2010, 164).

## 4. Method

The target items in this paper are four Japanese conditional markers: *to*, *ba*, *tara*, and *nara*. There are many rules to distinguish these, which are too complicated for learners. For example, if the conditional relationship is a permanent one (e.g., Push this button, and water splashes), *to* is used. However, to express feeling (e.g., If you push this button without permission, I will beat you), *to* cannot be used.

Before continuing the explanation, a definition of terms is necessary. A conditional expression is a sentence with any of the four markers above, regardless of its meaning. A conditional sentence refers to a relationship where if p is true, q is also true, and if p is false, q is also false. This is what traditional grammars mean by the term "conditional sentence." Please note that the interpretation of conditional sentences in natural language is different from the implication relation of propositional logic.

Furthermore, it is important to note that Japanese has no subjunctive mood as a grammatical category. This means that you cannot discern whether the sentence corresponds to a traditional conditional sentence from a specified form.

The data used in the paper came from the Balanced Corpus of Contemporary Written Japanese (BCCWJ). This corpus contains 100 million words and consists of 13 genres; therefore, it is balanced. The corpus also provides a web-based searching application *Chunagon* (https://chunagon.ninjal.ac.jp/), which enables users to search for morpheme information. Since the corpus allows the user to select a part of speech, it is possible to pick out verbs that occur before specific markers. This study searched the corpus with two conditions for each marker: a verb that appears before a conditional marker and a verb that appears before another function word before a conditional marker.

The number of example sentences for each marker is shown in Table 1 below.

*Table 1: Number of example sentences for each marker*

| *To* | *Ba* | *tara* | *nara* |
|---|---|---|---|
| *219,995* | *135,634* | *87,201* | *16,504* |

# 5. Results

## 5.1 Verbs that appear before conditional markers

In the lexical approach, the first step is to create a table of verbs that frequently collocate with the target item.

Firstly, verbs that combine with *nara* follow a different pattern than verbs that combine with the other three markers. The reason for this difference is that *nara* is used mainly for counterfactual conditionals or to respond to others' utterances, a function that is not shared by the other conditional markers. This explains the different selection of verbs. In Table 2, the verbs, their frequency, ratio of frequency, the rank of token frequency of the verbs in the full corpus, and the meaning group, which is discussed in the next section, are shown.

*Table 2: Top 20 verbs that appear before* nara

| *Rank* | *Verb* | *Frequency* | *Ratio* | *Rank of the verb* | *Group* |
|---|---|---|---|---|---|
| 1 | *SAY (iu)* | *1,655* | 10.03% | 4 | *(B)* |
| 2 | *DO (suru)* | *1,584* | *9.60%* | 1 | |
| 3 | *BE-inanimate (aru)* | *1,110* | 6.73% | 3 | *(E)* |
| 4 | *BECOME (naru)* | *390* | 2.36% | 2 | *(C)* |
| 5 | *CAN (dekiru)* | *351* | 2.13% | 10 | |
| 6 | *THINK (omou)* | *347* | 2.10% | 5 | *(B)* |
| 7 | *GO (iku)* | *329* | 1.99% | 8 | *(D)* |
| 8 | *THINK (kanagaeru)* | *310* | 1.88% | 11 | *(C)* |
| 9 | *SEE (miru)* | *263* | 1.59% | 6 | *(A)* |
| 10 | *BE-animate (iru)* | *257* | 1.56% | 9 | *(E)* |
| 11 | *DO-casual (yaru)* | *226* | 1.37% | 16 | |
| 12 | *USE (tsukau)* | *166* | 1.01% | 20 | |
| 13 | *COME (kuru)* | *138* | 0.84% | 7 | *(D)* |
| 14 | *GO OUT (deru)* | *135* | 0.82% | 14 | *(D)* |
| 15 | *BUY (kau)* | *120* | 0.73% | 40 | |
| 16 | *UNDERSTAND (wakaru)* | *116* | 0.70% | 13 | *(A)* |
| 17 | *TAKE (toru)* | 93 | 0.56% | 22 | |
| 18 | *HOPE (nozomu)* | *90* | 0.55% | 303 | |
| 19 | *KNOW (shiru)* | 89 | 0.54% | 18 | *(C)* |
| 20 | *DESIRE (motomeru)* | 81 | 0.49% | 48 | |

The verbs ranked from 1–10 also rank highly in the full corpus data; the verbs that are not used frequently in the full corpus are ranked 11–20. It is necessary to stress that most of the examples are counterfactual conditionals.

In contrast, the other three markers are rarely used for counterfactual conditions and show similar verb patterns. Table 3 presents the top 20 verbs that appear before *to*.

*Table 3: Top 20 verbs that appear before* to

| *Rank* | *Verb* | *Frequency* | *Ratio* | *Rank of the verb* | *Group* |
|---|---|---|---|---|---|
| 1 | *SEE (miru)* | 20,304 | *9.23%* | 6 | *(A)* |
| 2 | *DO (suru)* | 17,304 | *7.87%* | 1 | |
| 3 | *BECOME (naru)* | 16,688 | *7.59%* | 2 | *(C)* |
| 4 | *SAY (aru)* | 15,184 | *6.90%* | 4 | *(B)* |
| 5 | *THINK (kanagaeru)* | 5,172 | *2.35%* | 5 | *(B)* |
| 6 | *GO (iku)* | 5,013 | *2.28%* | 8 | *(D)* |
| 7 | *HEAR (kiku)* | 3,519 | *1.60%* | 17 | *(A)* |
| 8 | *THINK (omou)* | 3,030 | *1.38%* | 11 | *(B)* |
| 9 | *COMPARE (kuraberu)* | 2,792 | *1.27%* | 89 | *(A)* |
| 10 | *COME (kuru)* | 2,775 | *1.26%* | 7 | *(D)* |
| 11 | *GO IN (hairu)* | 2,703 | *1.23%* | 19 | *(D)* |
| 12 | *GO OUT (deru)* | 1,873 | *0.85%* | 14 | *(D)* |
| 13 | *BE-animate (iru)* | 1,546 | *0.70%* | 9 | *(E)* |
| 14 | *BE-inanimate (aru)* | 1,447 | *0.66%* | 3 | *(E)* |
| 15 | *PUT IN(ireru)* | 1,431 | *0.65%* | 27 | *(D)* |
| 16 | *DO-casual (yaru)* | 1,413 | *0.64%* | 16 | |
| 17 | *USE (tsukau)* | 1,283 | *0.58%* | 20 | |
| 18 | *TAKE (toru)* | 1,276 | *0.58%* | 22 | |
| 19 | *GIVE (ageru)* | 1,146 | *0.52%* | 29 | |
| 20 | *ARRIVE (tsuku)* | 1,020 | *0.46%* | 25 | *(D)* |

As many as 10% of conditional sentences contain “SEE,” which appears to contradict our intuition. Furthermore, “COMPARE” is not a very frequent word; it ranks 89th in the corpus, but reaches 9th position before conditional markers.

*Table 4: Top 20 verbs that appear before* ba

| *Rank* | *Verb* | *Frequency* | *Ratio* | *Rank of the verb* | *Group* |
|---|---|---|---|---|---|
| 1 | *SAY (iu)* | *19,171* | *14.13%* | 4 | *(B)* |
| 2 | *DO (suru)* | *12,638* | *9.32%* | 1 | |
| 3 | *BE-inanimate (aru)* | *10,178* | *7.50%* | 3 | *(E)* |
| 4 | *BECOME (naru)* | *6,151* | *4.53%* | 2 | *(C)* |
| 5 | *SEE (miru)* | *4,287* | *3.16%* | 6 | *(A)* |
| 6 | *CAN (dekiru)* | *3,357* | *2.48%* | 10 | |
| 7 | *GO (iku)* | *2,988* | *2.20%* | 8 | *(D)* |
| 8 | *THINK (kanagaeru)* | *2,619* | *1.93%* | 11 | *(B)* |
| 9 | *THINK (omou)* | *2,350* | *1.73%* | 5 | *(B)* |
| 10 | *BE-animate (iru)* | *2,236* | *1.65%* | 9 | *(E)* |
| 11 | *COME (kuru)* | *1,281* | *0.94%* | 7 | *(D)* |
| 12 | *DO-casual (yaru)* | *1,260* | *0.93%* | 16 | |
| 13 | *USE (tsukau)* | *1,147* | *0.85%* | 20 | |
| 14 | *ACCORD (yoru)* | *949* | *0.70%* | 78 | *(A)* |
| 15 | *HEAR (kiku)* | *900* | *0.66%* | 17 | *(A)* |
| 16 | *COMPARE (kuraberu)* | *879* | *0.65%* | 16 | *(A)* |
| 17 | *UNDERSTAND (wakaru)* | *859* | *0.63%* | 89 | *(A)* |
| 18 | *TAKE (toru)* | *688* | *0.51%* | 22 | |
| 19 | *SAY IN OTHER WORDS (iikaeru)* | *686* | *0.51%* | 1,066 | *(B)* |
| 20 | *GO IN (hairu)* | *678* | *0.50%* | 19 | *(D)* |

The 14th most frequently occurring word in front of *ba* is "ACCORD," which is used to say "according to something" in Japanese. This expression consists of a verb + conditional marker. Even more noteworthy is the verb "SAY IN OTHER WORDS," ranked 19th, which is not very frequent in the corpus.

*Table 5: Top 20 verbs that appear before* tara

| *Rank* | *Verb* | *Frequency* | *Ratio* | *Rank of the verb* | *Group* |
|---|---|---|---|---|---|
| 1 | *DO (suru)* | *10,699* | *12.27%* | 1 | |
| 2 | *BECOME (naru)* | *5,042* | *5.78%* | 2 | *(C)* |
| 3 | *SAY (iu)* | *3,993* | *4.58%* | 4 | *(B)* |
| 4 | *BE-inanimate (aru)* | *3,978* | *4.56%* | 3 | *(E)* |
| 5 | *THINK (omou)* | *3,066* | *3.52%* | 5 | *(B)* |
| 6 | *SEE (miru)* | *2,625* | 3.01% | 6 | *(A)* |
| 7 | *COME (kuru)* | *2,067* | *2.37%* | 7 | *(D)* |
| 8 | *BE-animate (iru)* | *1,706* | *1.96%* | 9 | *(E)* |
| 9 | *GO (iku)* | *1,686* | *1.93%* | 8 | *(D)* |
| 10 | *HEAR (kiku)* | *1,373* | *1.57%* | 17 | *(A)* |
| 11 | *DO-casual (yaru)* | *1,167* | *1.34%* | 16 | |
| 12 | *CAN (dekiru)* | *1,028* | *1.18%* | 10 | |
| 13 | *GO OUT (deru)* | *914* | *1.05%* | 14 | *(D)* |
| 14 | *FINISH (owaru)* | *761* | *0.87%* | 62 | *(C)* |
| 15 | *THINK (kangaeru)* | *729* | *0.84%* | 11 | *(B)* |
| 16 | *ARRIVE (tsuku)* | *645* | *0.74%* | 25 | *(D)* |
| 17 | *GO IN (hairu)* | *574* | *0.66%* | 19 | *(D)* |
| 18 | *PROBABLY(hyottosuru)* | *549* | *0.63%* | 933 | |
| 19 | *PUT IN (ireru)* | *489* | *0.56%* | 27 | *(D)* |
| 20 | *TAKE (toru)* | *454* | *0.52%* | 22 | *(D)* |

The phrase *hyottoshitara* is ranked 18th. This verb is used only in this form and only in conditional sentences.

## 5.2 Groups of verbs

This section shows that most of the verbs that appear before the three markers *to*, *ba*, and *tara* fall into five groups based on meaning. These groups are indicated in the rightmost column in the table. The verb "DO" is an exception because it can be used to substitute for other verbs, like the English verb *do*, since it can convey several meanings.

**(A) Verbs of perception or information processing**

The first group contains verbs of perception or information processing. "SEE" or "HEAR" are verbs of perception and "COMPARE" or "ACCORD" are verbs of processing information. At first sight, they may fall into different groups, but examples of "SEE" or "HEAR" involve perception not of real objects but of information sources, such as graphs or news, whose contents follow. Thus, perception and information processing verbs can be in the same group. Below are two examples.

(3) *Kono hyou=wo mi=tara sugu wakar-u.*
this table=ACC see-COND soon understand-PRES
'This table is easy enough to understand just by looking at it.'

(4) *Houkoku=ni yoru=to, 5,000-nin=ga sanakashi-ta.*
report=DAT accord-COND 5,000-QUANT=NOM join-PAST 'According to the report, 5,000 people joined.'

As is expressed with participial constructions in the translation of English, these sentences are not used for conditional sentences in traditional meaning, despite conditional expressions being used.

**(B) Verbs of saying or thinking**

The second group is verbs of saying and thinking, such as "SAY" or "THINK." These verbs are usually followed by a complement of quotation. Combined with conditional expressions, they fall into two patterns. Consider the examples below.

(5) *Taro ga "yaa" to iu=to, Hanako=wa hohoen-da.*
Taro=NOM hi COMP say=COND Hanako=TOP smile-PAST
'Taro said "Hi," and Hanako smiled.'

(6) *Kantan-ni ie=ba, sou da.*
Simple–ly say=COND so COP
'To put it simply, that's right.'

The first pattern is an utterance followed by another person's utterance or action, as shown in example (5). In Japanese, when you conjoin two actions of speaking, you must use conditional markers, which is a difficult rule for learners. The second pattern is used for adverbial phrases like "Simply saying," "Strictly speaking," "In other words," and so on. These can be uttered as meta-discourse markers that show the function of the main clause. On this point, this pattern is similar to group (A), which shows the information source of the main clause. It is remarkable that both patterns do not express conditional meaning, although the verbs are conditional-marked.

**(C) Verbs of time**

The third group contains "BECOME" and "FINISH." Why do these verbs belong in the same group? The following example sentences illustrate the reason.

(7) *Haru=ni nat=tara, sakura=ga kirei=da.*
spring=DAT become=COND cherry.blossom=NOM beautiful=COP
'Cherry blossoms are beautiful in the spring.'

(8) *Shigoto=ga owat=tara, biiru=ga nomi=tai.*
work=NOM finish-COND beer=NOM drink-want.to
'I want to have beer after finishing work.'

The translation above shows that the sentences contain expressions of time such as 'in the spring' or 'after finishing work.' It is these parts that are expressed in conditional expressions in Japanese. For this reason, this group is named "verbs of time." The verb "BECOME" is used very widely in the language in general, but most conditional-marked instances concern time. Again, sentences with these conditional-marked verbs are not conditional sentences in the traditional sense.

**(D) Verbs of movement**

The fourth group is verbs of movement. This group contains "GO," "COME," "GO IN," "GO OUT," and "ARRIVE." "Put into" is included in this group because when you put a book into a bag, the book undergoes a movement. Conditional expressions with these verbs mean "If you move, you can see something." In other words, the meaning of this group is "finding after moving." Below are two examples.

(9) *Massugu ike=ba, toire=ga aru.*
straight go=COND toilet=NOM be
'If you go straight, there is a toilet.'

(10) *Soto=ni deru=to, ame=ga hut-tei-ta.*
Outside=DAT go.out=COND rain=NOM fall-PROG-PAST
'When I went out, I found it was raining.'

Examples (9) and (10) represent the typical pattern of this group. It can be said that conditional expressions with these verbs are used for conditional sentences in a traditional sense, unlike the three groups introduced previously, albeit in a different sense, because the sentences are not hypothetical or counterfactual, but factual.

**(E) Verbs of existence**

The fifth and final group contains verbs of existence, that is, verbs that translate as 'be' or sometimes as 'have.' Two examples are shown below.

(11) *100-man-en at=tara, nani=ga shi=tai?*
1-million-yen be=COND what=NOM do=want.to
'If you had one million yen, what would you want to do?'

(12) *Taro=ga i=tara, yokat=ta=noni!*
Taro=NOM be-COND good=PAST=though
'Wish Taro were here!'

These sentences illustrate that the verbs are used for the conditional meaning in the typical, classical sense.

## 6. Discussion

As observed in the previous section, verbs that appear in conditional expressions fall into five groups. One crucial observation is that only one of the five groups is used for "typical" conditional sentences, such as "If I had money, I'd go to Japan on vacation." This result of the corpus survey questions why such sentences are treated as "typical," since many of the instances are not of this type. This section examines the question "what is a conditional expression?" based on the results of the corpus survey.

English conditional sentences are known to fall into three subcategories: predictive or open conditionals, counterfactual or hypothetical conditionals, and epistemic conditionals (Kaufman 2001, Quirk et al. 1985). In Japanese, Arita (2005) inherits the categorization of English, and Maeda (2009) proposes four subcategories: potential conditionals, counterfactual conditionals, generic conditionals, and factual conditionals. Potential conditionals correspond to predictive conditionals such as *if it rains tomorrow, I will not go out*. Generic conditionals express a permanent relationship, such as *if you add up all the interior angles of a triangle, the result will be 180 degrees*. An example of factual conditionals is, *When I got home, I found that somebody was in my room*, which has no corresponding conditional structure in English.

However, the boundaries of these categorizations are so complicated that judgment is difficult even for researchers, so, of course, they are too difficult for learners to utilize. For example, although both Arita (2005) and Maeda (2009) categorized counterfactual conditionals, Japanese language does not have mood (e.g., subjunctive mood) as a grammatical category. It is not possible to discern whether a sentence has a counterfactual meaning by examining its form. This categorization is merely an inheritance from English grammar and does not benefit learners.

The lexical approach, which concentrates on verb meanings, provides a different illustration. In this approach, there exist mainly five categories of conditional expressions:

(A) Resources of information or medium of communication
(B) Events after saying or thinking
(C) Events that occur at a particular time
(D) Exploring something after moving
(E) Events that occur when hypothesizing that something exists

(E) is what is traditionally called a "conditional sentence." However, conditional expressions in Japanese have four other categories, (A) through (D), that are widely used. These categorizations easily connect with situations in which conditional markers are used and are helpful for learners to imagine using markers in daily life.

From the point of view of the lexical approach, sentences with conditional markers in Japanese can be summarized as in Fig.1.

*Figure 1. Summary of conditionals from the lexical approach.*

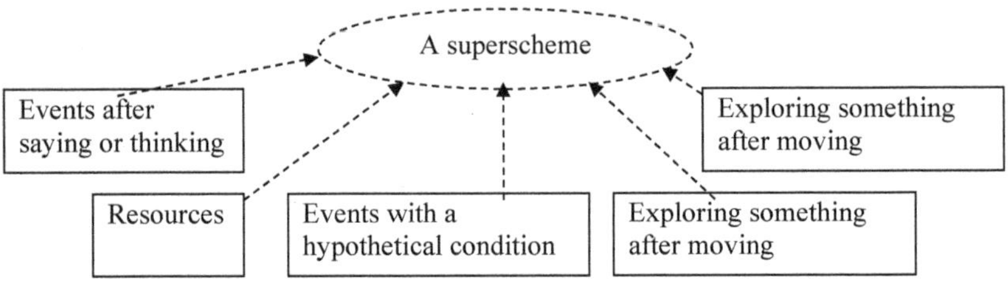


There are five subcategories of conditional expressions, so a "condition" is a superscheme rendered from the five subcategories.

Following this, the next question is, "what is the superscheme?" The superscheme can be seen in light of mental space theory, where it would be considered a space-introducing expression. However, it is unnecessary to describe the superscheme in language teaching. All learners require are the five subschemes and knowledge regarding when to use them. Each category is easily accessible to learners. For example, "Events after saying" is very different from a traditional conditional sentence, so learners tend to fail to use the conditional marker to express "Taro said 'Hi,' and Hanako waved her hand" and alternatively connect the two events with "AND," which is ungrammatical. To help learners, we must abandon the traditional idea that "conditional expressions express the same meaning as conditional sentences in traditional grammar express" and adopt the new user-friendly categorization. This categorization is more usable in a functional or topic based syllabus than the traditional categorization.

Factual and hypothetical conditionals still need to be distinguished, but this difference can be explained by difference of tense (e.g., *Turn right, and you will find a toilet.* vs. *I turned right and found a toilet*).

## 6. Conclusion

In this paper, the *Lexical Approach* is discussed through a corpus-driven survey concentrating on collocations of Japanese conditional markers. In conclusion, the corpus data do not support the presupposition that conditional markers are specialized for conditional sentences. The paper shows that Japanese conditional expressions fall into five subschemes according to the meaning of the verbs. The lexical approach thus provides not only word lists for teachers, but also new insights for language studies.

## References

*Arita, S.* (2007). *Nihongo-Jouken-Bun to Jiseisetsusei.* [Japanese Conditionals and Tensedness.] Tokyo: Kuroshio Shuppan.

*Bybee, J.* (2010). *Language, Usage and Cognition,.* Cambridge: Cambridge University Press.

*Kaufman, S.* (2001). Aspects of the meaning and use of conditionals. PhD Dissertation, Stanford University.

*Lewis, M.* (2000). There is nothing as practical as a good theory. Lewis, M. (Ed.) *Teaching Collocation: Further Development in the Lexical Approach.* pp.10–27. Heinle: Cengage Learning.

*Maeda, N.* (2009). *Nihongo no Fukubun.* [Complex Sentences in Japanese.] Tokyo: Kuroshio Shuppan.

*Nakamata, N.* (2011). Corpus-Driven Approach ni yoru Nihongo-Kyouiku-Bumpou-Kenkyuu: 'Tearu' to 'Teoku' wo Rei to shite. [A Study of Japanese Pedagogical Grammar by Corpus Driven Approach: In the case of 'Tearu' and 'Teoku.'] Mori, A., & Iori, I. (Eds.) *Nihongo-Kyouiku-Bumpou no tame no Tayouna Approach.* [Various Approaches for Japanese Pedagogical Grammar.] pp. 215–233. Tokyo: Kuroshio Shuppan.

*Quirk, R., et al.* (1985). *A comprehensive grammar of the English language*. Edinburgh: Pearson Education.

*Woodland, G.* (2000). Collocation–encouraging learner independence. Lewis, M. (Ed.) *Teaching Collocation: Further Developments in the Lexical Approach.* pp. 28–46. Heinle: Cengage Learning.

Piotr Twardzisz
*University of Warsaw, Poland*

# Identifying and measuring personification in journalistic discourse

**Abstract:** This article reports on the results of research that examines the degree of the metaphoricity of state names as they appear in press texts on international relations (IR). More specifically, the degree of the personification of these names is thoroughly examined. Initially, a rigorous criterion for establishing what constitutes a state name has to be assumed. Consequently, the search involves all the names of the 192 members of the United Nations (as of February 2011). Our search covers all possible (grammatical) patterns of use as well as the linguistic behaviours of the widest possible range of names. The results of this investigation do not confirm the widely held belief about the ubiquity and pervasiveness of the "state is a person" metaphor.

Keywords: metaphor, personification, state name, quantitative and qualitative analysis

## 1. Introduction

The aim of the study reported on in this article[1] is to establish the realistic degree of metaphoricity among proper names in the discourse of international relations. Our linguistic search focusing on state names will concern their alleged ubiquitous personification. The choice of journalistic language for this purpose may seem deliberate, as this type of language is commonly believed to be more florid than sober and unembellished scientific discourses (cf. Croft & Cruse, 2005). However, this assumption is also commonly rejected in the literature (see, e.g., Nerlich & Koteyko, 2009, p. 156). In their detailed study on metaphor identification, Steen et al. (2010) have established that 79.9% of words in the news register are non-metaphorical. As little as 15.0% of the lexical material analysed has unanimously been coded as metaphorical by the four coders involved. Overall, Steen et al. (2010) do not confirm the high frequencies of novel language in their news corpus, though they do not preclude a higher level of metaphorical occurrences in some sub-genres of news texts. The results of our empirical research are even less supportive of the common view that journalistic language is metaphorical.

1 This article is an abridged and modified version of Chapter 5 of Twardzisz (2013). Reproduced with permission of Palgrave Macmillan.

## 2. The language of international relations (IR)

In this research, a choice has been made to approach IR language aimed at a "general" audience, which naturally may include both specialists and non-specialists in international affairs. Given this, academic publications, scholarly books and articles, and specialist reports relating to any issues under international politics have been excluded from our analysis, as such sources are more "specialist". It is the IR language reaching the so-called general audience that this research has focused on, though we need to keep in mind the fact that the general audience potentially includes everyone. For instance, aiming at metaphor identification, Steen et al. (2010) mention "metaphor to the general language user" (p. 34) and the assumption of the existence of a "general user" (p. 112) with "general world knowledge" (p. 44). In their analysis of metaphorical economic language, Silaški and Đurović (2010) emphasize the importance of "popular economic discourse" (p. 57), as opposed to specialist economic discourse, as the source of metaphorical expressions.

In practice, it is difficult to establish a neatly delineated set of texts aimed at specialists. Chilton and Lakoff (1995) point out that there are differences between "folk" (p. 37) concepts relating to world affairs and those shared by specialists in the field of IR. However, these concepts and discourses can hardly be separated from each other. Not only can both kinds of audiences use both types of texts, but mixing of knowledge between the two types of audiences is also on the increase, diminishing the pre-existing divide. Therefore, Chilton and Lakoff (1995) assume "some degree of continuity between the specialist and the everyday discourse of foreign policy" (pp. 37–38) in their account of metaphors in foreign policy. It is doubtful, though, whether any texts dealing with world affairs are ever approached by people who have no background knowledge whatsoever of foreign policy, international security issues, or international trade. Thus, in our research, we will rely on texts that do not necessarily cater for trained IR specialists, but for readers who come from different walks of life and are knowledgeable about world affairs, no matter where this knowledge comes from.

Having reduced the influence of academic and other highly specialist language sources, we will concentrate our attention on the language of the media. Being more "public", the language of the media differs from more "institutional" discourses (e.g., legal, medical, or technical) in that it addresses more general and broader audiences (cf. Lauerbach & Fetzer, 2007, pp. 14–15). It is impossible not to be influenced by the media with respect to world affairs, as for most people the media, including the internet, are the only source of information about the world. In practice, only a handful of highly trained specialists are in a position to seek and acquire information, and have their views about IR shaped, without being influenced by the media (cf. Hill, 2003, p. 274).

## 3. Empirical methods employed in the analysis

We rely exclusively on the written mode to the exclusion of the spoken and visual modes. Our assumption is that the written language of IR found in press coverage is linguistically more relevant to our study than the elusive spoken or imagistic language of non-print media. There may, however, be the danger of deliberate manipulation in favour of metaphor production in written language, as journalists have more time to "carefully craft their texts and make precise lexical choices" (see Biber, 1988, pp. 104–105; after Steen et al., 2010, pp. 43–44).

We have decided to scour the language of weekly magazines rather than daily newspapers. Daily newspapers, though often rich in IR content, would be too difficult to manage in a broader time frame due to the number of issues to consider. The decision concerning the choice of weekly magazines has a practical dimension. The prototypical recipient of IR press language has enough time to access one or two weeklies with their rich IR content per week, rather than every daily issue of a few quality newspapers with their bulging foreign affairs sections. In the case of daily newspapers, it would be manageable to search a few titles within a month. Such a short time frame would perhaps limit and distort the results.

The availability of weekly magazines to a more general public has outweighed limited accessibility to language corpora. The search undertaken here resembles as closely as possible the reading process of major international weekly news magazines by an educated reader interested in world affairs. The data collected in our search are meant to resemble the linguistic stock that possibly shapes the world views of a real living reader, rather than an occasional linguist–analyst.

In order to avoid too much weighting on any one particular variety of English, two internationally renowned weekly magazines published on either side of the Atlantic have been selected: *Newsweek* and *The Economist*. In order to limit the enormous amounts of data involved, one American and one British weekly have been considered sufficient. Our ultimate goal has not been to compare and/or contrast the languages and contents of both weeklies. There are numerous other studies comparing different framings and viewpoints of selected phenomena (cf. Abrahamian, 2003; Esser & D'Angelo, 2006; Ruigrok & van Atteveldt, 2007; Esser, 2008; Papacharissi & Oliveira, 2008).

All the 192 state names have been divided into five frequency groups. The groups are: the one-digit group, two-digit group, three-digit group, four-digit group, and five-digit group. Membership of a given state name in one of these groups depends on the average number of occurrences of this name in three magazines: *Newsweek*, *Time*, and *The Economist*. The search periods for these three magazines mostly overlap and are: "any time" (*Newsweek*), "from 1997 to 2011" (*Time*), and "from 1997 to 2011" (*The Economist*). The period of 14 years from 1997 to 2010 is sufficiently long to obtain a more balanced frequency than that arrived at in an arbitrarily selected year. Quantitative data collected in a more extended time frame guarantee less fluctuation caused by frequent temporary vicissitudes of usage of particular names.

The average number of occurrences has been calculated by adding the three results (="all results") from the three magazines and dividing them by three. If the average number of results within the search period is a one-digit figure, this state (name) belongs to the one-digit group. If it is a two-digit figure, the state belongs to the two-digit group, and so on. The frequency disproportions of some of the names are surprising, ranging from 0 to 47,698.

As for the search period, it is the "past 12 months" option for both target magazines, that is *Newsweek* and *The Economist*. The data come from all printed issues selectively retrievable from the two magazines' online archives freely available in December 2010/January 2011, when the search was carried out. So, in most cases, the language data come from the 12-month search period between January 2010 and December 2010. The data extracted were visually scanned with particular focus on the key state names. All uses of state names were recorded and classified accordingly.

Though we have taken a wide-angle look at how the 192 names have been used in written texts, our primary concern has been the identification of any linguistic instantiations of the 'state is a person' metaphor. Our analysis is also meant to fill the apparent gap to which Chilton (1996) alludes when he writes about the shortage of explicit accounts of "the way metaphor works" and "detailed textual analyses of occurrences of metaphor" (p. 38). The rigorous linguistic approach assumed here also meets the requirements laid out in Steen et al. (2010) when they write of researchers' obligation to demonstrate real language data as "independent entities in the world out there" (p. 149). Having collected and catalogued relevant language data, we want to be able to estimate whether the amount of metaphorical language identified justifies common assumptions such as "[t]he *heavy* presence of such and similar expressions" (Kövecses, 2009, pp. 20–21; italics added).

Our methodology is based on a three-stage selection process. Initially, "all results" are established on the basis of individual searches targeting state names in the two weekly magazines in the 12-month time period. Next, out of "all results", any relevant uses of all state names corresponding to "target constructions" are listed. There are five such "target constructions" in which a state name participates: subject of an active sentence, subject of a passive sentence, object of an active sentence, object in an optional "*by*-phrase" of a passive sentence, and the Saxon genitive construction. Finally, "possible personifications" are separated from "non-personifications" on the basis of a careful qualitative analysis of the "target constructions". Establishing "possible personifications" is burdened with some subjectivity, as most of the occurrences in the "target constructions" group can be understood by different analysts as potential carriers of human-like traits. Thus, in each individual case the analyst answers the question whether a given expression conforming to one of the target constructions not only reveals a human-like quality, but confers this quality onto the state name involved. In most cases, the search for a human-like quality focuses on the neighbouring word, or a few words if necessary. If these neighbouring words do not reveal any human-like qualities, a given expression is excluded from "possible personifications". Also, if these neighbouring words do not confer a human-like quality on the key state name, this occurrence is excluded from "possible personifications", and is classified as a "non-personification". The notion of personification as invoked in "non-personifications" and "possible personifications" carries a rather loose sense of the term. At this stage in the discussion of the "state is a person" metaphor, the use of "personification" has a maximally broad sense which refers to any human-like trait perceivable in a given state name, be it metaphorical, metonymic, or something else.

## 4. Results of the analysis

The majority of "all results" constitute non-relevant cases. Generally, typical locative expressions in the form of prepositional phrases have been excluded. State names used attributively in complex noun phrases (e.g., *the Yemen regime is worried*) and in descriptive phrases (e.g., *King Abdullah of Jordan)* have been avoided. Various irrelevant occurrences, such as *The Bank of America, Air France, New Mexico, Michael Jordan, an Iraq war veteran, Georgia* (a US state), and so forth, had to be removed from the pool of relevant results.

From among the five target grammatical constructions under consideration, the subject of an active sentence poses a challenge for establishing "possible personifications". For such constructions, we have applied the following procedure. If a state name is followed by an intransitive verb with an unequivocally human sense, we mark this occurrence as a possible personification (e.g., *Syria repented, Hungary winked*, etc.). Similarly, if a state name is followed by a transitive verb with a potential human sense and this is followed by an object whose sense, combined with that of the preceding verb, results in a human sense of the state name, we also mark this occurrence as a possible personification (e.g., *France devoted a lot of attention*, *Spain spent little time over*, etc.). However, if a given name is followed by a transitive verb with a potential human sense and it is followed by an object whose sense, combined with that of the preceding verb, does not result in a human sense of the state name, we do not mark this occurrence as a possible personification (e.g., *Turkmenistan developed its gas field*, *Vietnam ordered six submarines*, etc.). What seems to be at issue in both types of examples above is that the nominals corresponding to the direct objects either designate (*attention* and *little time*) or do not designate (*gas field* and *submarines*) entities normally associated with individual humans. The former can be easily imagined in all kinds of contexts where individuals play a major role, while the latter are thought of as characterizing groups, institutions, companies, and so forth. In many cases, occurrences with seemingly the same verb have had to be classified differently. For example, *Japan tightened its structural deficit* has been marked as a case of non-personification, but *Japan tightened its belt* as possible personification.

We have assumed sedimented or dead metaphors as non-personifications in our selection process. Expressions such as *Ireland grappled with its banking crisis* or *Iran grabbed everyone's attention* have been marked as non-personifications due to their sedimentation of metaphoricity. In this, we have followed Black (1993, p. 25), who considers sedimented or dead metaphors such as *France allied with Germany* as non-metaphorical (after Drulák, 2006, p. 507). Ambiguous cases such as *France's retirement age* and *France's remarkable fertility* have been excluded from possible personifications. Their contextual uses indicate the indirect reference to "people living in France" rather than the state per se. Such occurrences have been treated as less clear but shorter ways of phrasing more wordy statements such as *the retirement age of the people living in France* and *the remarkable fertility of the people living in France*, respectively.

The following five sections present detailed results of the quantitative and qualitative analysis of the linguistic behaviour of the key state names in the five target constructions in the 12-month time period. The language data obtained in the search have undergone the three-stage selection process establishing a pool of possible personifications for each frequency group.

## 4.1 The one-digit group

This group includes eight names. In our 12-month search of *Newsweek* and *The Economist*, one of these names has been found once only in "all results", which also happens to be a "relevant result". One relevant occurrence of *Antigua and*

*Barbuda* has been recorded: *Antigua and Barbuda has also turned to Canada....* No other names in this group have been recorded in the 12-month search period. With only one occurrence of one of these names, there is no evidence to uphold *any* metaphoricity of the state at this stage.

## 4.2 The two-digit group

This group consists of 51 names of UN members. There has been no single occurrence of any of the 51 names in a *by*-phrase in a passive sentence. Also, the state name as the subject of a passive voice sentence is a rare phenomenon here. Similarly, only ten names have been recorded in the position of an object in an active sentence. The only instance exhibiting a human-like quality is *threaten Burundi*. No more than 50% of the names in this group have been found in the Saxon genitive. In the vast majority of the recorded uses, the genitive constructions appear in political or economic, domestic rather than international, contexts (e.g., *Lesotho's election*, *Equatorial Guinea's prisons*, etc.).

Thirty-six out of 51 names have been recorded in the subject position. Stative processes expressed by means of copula verbs are the most prominent as a group. Among the above copular and stative constructions, only one instance of a "possible personification" has been recorded, namely *be bilingual*. Approximately a third of the other occurrences display some dynamicity due to a unique combination of a state name with a dynamic verb.

The majority of these names in the subject position have been marked as non-personifications. Instances such as *allow in a trickle of tourists* have been marked as non-personifications because they represent combinations of a potentially human verb with an object whose semantic scope is much too broad for an individual human subject. The nominal (*a*) *trickle of tourists* corresponds to an entity larger than an individual person is capable of handling.

Sixteen occurrences of state names in the position of the subject of an active sentence have been marked as possible personifications. Those personifications are accompanied by predicates such as: *allow someone to do something*, *need stable neighbours*, *need cash*, *want to learn*, and so forth. The assumption is that the verbs and their objects form an appropriate semantic environment for the respective state names to possibly correspond to individual humans. These 16 occurrences and four possible personifications identified in other constructions (*threaten Burundi*, *Guyana's neighbours*, *Kyrgyzstan's approval*, and *Malta be bilingual*) make 20 possible personifications in this group. This number constitutes 5.35% of 374 "all results" retrieved in this group.

## 4.3 The three-digit group

The search of the 96 names in this group has produced 5,048 "all results" from both sources combined in the 12-month time period. The two names with the highest "all results" are *Greece* (363) and *Portugal* (177), and those with the lowest "all results" are *Bahamas* (7) and *Paraguay* (9). The vast majority of the remaining 92 names have been found in fewer than 50 "all results". Only some of those have been found in more than 50 "all results".

Passive statements with an optional state name in a *by*-phrase are not popular in this group. Only 12 (12.5%) names have occurred 14 times, nine of which have been tentatively marked as possible personifications. Approximately a third of the names in this group have been recorded in the position of the subject of a passive sentence. Only two occurrences have been marked as possible personifications, namely, (*Zimbabwe*) *be accused* (*of torture*) and (*Haiti*) *be decapitated*. Most (70, i.e., 73%) names have been found in the position of the object of an active sentence, totalling 186 occurrences, which changes the trend in the use of state names observed so far. One hundred and thirty-four occurrences in the "object" search have been selected as non-personifications. The remaining 52 occurrences (almost 28% of all objects) of names in the object position have been classified as possible personifications (e.g., *blame Ukraine*, *scar Haiti*, etc.). Altogether, there are 581 occurrences in the Saxon genitive. Initially, we have put aside 526 (approx. 90.5%) instances which do not reveal any human-like traits as non-personifications. Fifty-five (approx. 9.5%) out of 581 occurrences in the Saxon genitive have been marked as possible personifications (e.g., *Iceland's cousins in Norway*, *Sri Lanka's experience*, *Yemen's bad habits*, etc.).

The last construction to consider in this group is a state name as the subject of an active voice sentence. Altogether, we have found 764 occurrences, the vast majority of which (597, approx. 78%) have been regarded as non-personifications. One hundred and sixty-seven (that is, 21.9%) out of 764 occurrences of names in the subject position have been marked as possible personifications. The following examples constitute a representative sample of those occurrences: (*Albania*) *feel* (*lonely*), (*Angola*) *suggest* (*a better solution*), (*Angola*) *potter along*, (*Bahrain*) *get* (*edgy*), (*Colombia*) *deliver* (*heavy blows*), (*Oman*) *have* (*plans*), (*Sri Lanka*) *repent*, and so on.

There are 1,603 occurrences of target constructions, constituting 31.75% of all results. From among the five target constructions, 285 possible personifications have been elicited. This figure constitutes approximately 5.7% of all results.

## 4.4 The four-digit group

The four-digit group consists of the names of 35 UN members, designating countries whose presence in the media is, understandably, high. The search has produced

12,830 "all results" from both sources combined in the 12-month time period. Only 16 names have been found in 33 optional *by*-phrases in passive sentences. Sixteen occurrences have been coded as non-personifications and 17 as possible personifications. Similarly, only 17 out 35 names in this group have been found in 45 occurrences in the position of the subject of a passive sentence. Most (32) of these have been marked as non-personifications, while 13 occurrences have been singled out as "possible personifications" (e.g., *Sweden is confused).* In the position of the object of an active sentence, there were 234 occurrences recorded. One hundred and seventy-three occurrences (i.e., 74%) have been marked as non-personifications. Sixty-one occurrences (26%) have been assumed to be possible personifications (e.g., *mock Switzerland*). Among the Saxon genitive constructions retrieved in our search, the prevailing majority of occurrences are in geographical, judicial, economic, and sociopolitical contexts, with a mix of proper names. Altogether, 595 occurrences have been recorded, out of which 525 (i.e., 88%) have been marked as non-personifications. The remaining 70 occurrences (12%) have been tentatively classified as possible personifications (e.g., *Israel's loyal friend*).

Names in the subject position in active voice sentences are, understandably, frequent. The classification and division of all 1,063 multifarious occurrences into those marked as non-personifications and those labelled "possible personifications" has proved quite a challenge. Seven hundred and forty-one (i.e., 70%) instances have been analysed as non-personifications. Three hundred and twenty-two (i.e., 30%) out of 741 active voice subject occurrences have attracted our attention as possible personifications. Some clearly exhibit more mental types of human-like activities, as in: *Ireland is flirting with a debt crisis, Spain became a passionate advocate*, and so on. Some, though seemingly fewer, occurrences display physical types of human-like activities, such as: *Poland has overcome hurdles, South Africa bent over backwards*, and so forth.

There are 1,970 occurrences of target constructions, constituting 15.35% of all results. From among the five target constructions we have elicited 483 possible personifications. This figure constitutes 3.76% of all results.

## 4.5 The five-digit group

The five-digit group includes two UN members: the UK and the US. Several alternative names for these countries have been used in the search, but their separate frequencies have been added to produce a total number of all results for the two states involved. The search for all the above names has produced 10,512 "all results" from both sources combined.

Twenty-four instances of *by*-phrases in passive sentences have been extracted. Seventeen of those have been marked as non-personifications. Seven of those uses have been considered as possible personifications. In the position of the subject of a passive voice sentence, only names designating the US have been found. Out of 15 occurrences, 11 have been understood as non-personifications. In the position of the object of an active voice sentence, we have recorded 57 occurrences. Forty-two of those have been considered as non-personifications. Fifteen instances have been granted the temporary status of possible personifications (e.g., *bother the UK*, *outperform the US*, etc.). Altogether, 534 instances of the Saxon genitive with both state names have been recorded during the search. Four hundred and seventy-eight of those occurrences have been marked as non-personifications. The remaining 56 cases have been classified as possible personifications (e.g., *Britain's/America's anger*, *favourite pastimes*, etc.).

The two names in the position of the subject of an active voice sentence are, understandably, well represented. Five hundred and seventy-eight occurrences of these names in the subject position have been recorded. Four hundred and thirty-eight instances (64 for the UK and 374 for the US) have been analysed as non-personifications. Occurrences of possible personifications of the two names have been marked 140 times (35 for *the UK* and 105 for *the US*).

There are 1,208 occurrences of any of the five target constructions, constituting 11.5% of all results. From among the five target constructions, 222 possible personifications have been extracted. This figure constitutes 2.11% of all results.

## 5. Concluding remarks

The frequencies of use among the 192 UN member names differ quite dramatically. They range from only one occurrence to a few thousand occurrences in the two sources in a 12-month period. In many cases, a country's size, its population, period of UN membership, or period of independence do not directly and automatically translate into an increased presence in press language on IR. Only internationally *significant* state-actors have made the lists in the four- and five-digit groups. Their international significance may have different causes and motivations, be they economic, political, cultural, or a mixture of these.

The results of the analysis inform us about the percentages of possible personifications of state names in press language on IR. The label "possible personification" should be treated as maximally inclusive, if not somewhat inflated. The category "possible personification" has been more of an approximation than a rigorously defined set. Besides, the label "possible personification" has always been applied in case of doubt. Given the widespread inclusiveness of our parameters, it is important to observe that the numbers of possible personifications are low by any standards, that is, 5.35% (the two-digit group), 5.7% (the three-digit group), 3.76% (the four-digit group), and 2.11% (the five-digit group).

Once an expression has been marked as a possible personification, it enters an all-inclusive set encompassing expressions with different degrees of personification. The pool of possible personifications hosts different kinds of expressions carrying human-like qualities conferred onto state names. The requirement of semantic tension for a given meaning to be recognized as metaphorical has been raised in the literature (see Charteris-Black, 2004, p. 37). Some sentences in our database exhibit little, if any, semantic tension. Others show more semantic tension, as they are innovative, creative, or original, as stipulated in De Landtsheer (2009, p. 65). In other words, the amount of semantic tension is never the same. This observation suggests that further refinements of our preliminary assumptions about possible personifications need to be sought and implemented.

These reservations are also indicative of the fact that the general category of possible personifications is preliminary and temporary, rather than conclusive. Furthermore, it may be assumed that not all cases classified as possible personifications are, indeed, personifications at all. If this assumption is true, the percentages of definite personifications will be even lower than those established above.

The results of our data-based analysis, exhibiting very low numbers of linguistic personifications, cast doubt on the dictum about the ubiquity and pervasiveness of the "state is a person" conceptual metaphor. In fact, empirical evidence shows that the personification of the political state is anything but ubiquitous and pervasive.

## References

*Abrahamian, E.* (2003). The US media, Huntington and September 11. *Third World Quarterly, 24*(3), 529–544.

*Biber, D.* (1988). *Variation across speech and writing.* Cambridge: Cambridge University Press.

*Black, M.* (1993). More about metaphor. In A. Ortony (Ed.), *Metaphor and thought* (2nd revised ed., pp. 19–41). Cambridge: Cambridge University Press.

*Charteris-Black, J.* (2004). *Corpus approaches to critical metaphor analysis.* Basingstoke: Palgrave Macmillan.

*Chilton, P.* (1996). *Security metaphors: Cold War discourse from containment to common house.* New York: Peter Lang.

*Chilton, P. & Lakoff, G.* (1995). Foreign policy by metaphor. In Ch. Schäffner & A. L. Wenden (Eds.), *Language and peace* (pp. 37–59). Aldershot: Dartmouth.

*Croft, W. & Cruse, D. A.* (2005). *Cognitive linguistics* (3rd printing). Cambridge: Cambridge University Press.

*De Landtsheer, Ch.* (2009). Collecting political meaning from the count of metaphor. In A. Musolff & J. Zinken (Eds.), *Metaphor and discourse* (pp. 59–78). Basingstoke: Palgrave Macmillan.

*Drulák, P.* (2006). Motion, container and equilibrium: Metaphors in the discourse about European integration. *European Journal of International Relations*, *12*(4), 499–531.

*Esser, F.* (2008). Dimensions of political news cultures: Sound bite and image bite news in France, Germany, Great Britain, and the United States. *The International Journal of Press/Politics*, *13*(4), 401–428.

*Esser, F. & D'Angelo, P.* (2006). Framing the press and publicity process in U.S., British, and German general election campaigns: A comparative study of metacoverage. *The Harvard International Journal of Press/Politics*, *11*(3), 44–66.

*Hill, Ch.* (2003). *The changing politics of foreign policy*. Basingstoke: Palgrave Macmillan.

*Kövecses, Z.* (2009). Metaphor, culture, and discourse: The pressure of coherence. In A. Musolff & J. Zinken (Eds.), *Metaphor and discourse* (pp. 11–24). Basingstoke: Palgrave Macmillan.

*Lauerbach, G. E. & Fetzer, A.* (2007). Political discourse in the media. Cross-cultural perspectives. In A. Fetzer & G. E. Lauerbach (Eds.), *Political discourse in the media. Cross-cultural perspectives* (pp. 3–28). Amsterdam & Philadelphia: John Benjamins.

*Nerlich, B. & Koteyko, N.* (2009). MRSA – portrait of a superbug: A media drama in three acts. In A. Musolff & J. Zinken (Eds.), *Metaphor and discourse* (pp. 153–169). Basingstoke: Palgrave Macmillan.

*Papacharissi, Z. & de Fatima Oliveira, M.* (2008). News frames terrorism: A comparative analysis of frames employed in terrorism coverage in U.S. and U.K. newspapers. *The International Journal of Press/Politics*, *13*(1), 52–74.

*Ruigrok, N. & van Atteveldt, W.* (2007). Global angling with a local angle: How U.S., British, and Dutch newspapers frame global and local terrorist attacks. *The Harvard International Journal of Press/Politics*, *12*(1), 68–90.

*Silaški, N. & Đurović, T.* (2010). CATCHING INFLATION BY THE TAIL – Animal metaphoric imagery in the conceptualisation of INFLATION in English. *Ibérica*, *20*, 57–80. Retrieved June 22, 2014, from http://www.aelfe.org/documents/03_20_Silaski.pdf.

*Steen, G. J., Dorst, A. G., Herrmann, J. B., Kaal A. A., Krennmayr, T. & Pasma, T.* (2010). *A method for linguistic metaphor identification: From MIP to MIPVU.* Amsterdam & Philadelphia: John Benjamins.

*Twardzisz, P.* (2013). *The language of interstate relations. In search of personification.* Basingstoke & New York: Palgrave Macmillan.

Jarosław Wiliński
*Siedlce University of Natural Sciences and Humanities, Poland*

# A covarying collexeme analysis of the verb *play* and the *manner* adjunct in the domain of soccer

**Abstract:** This paper adopts a constructional approach to grammatical structure and a corpus-based method that is specifically geared to investigating interactions between lexical items occurring in two different slots of a grammatical construction. The method, referred to as *covarying collexeme analysis*, is used to identify the strength of association between the verb *grać* (*play*) and the manner adjunct occurring in the *act-manner* construction in Polish. The paper shows that there are pairs of lexical items that co-occur significantly more frequently than expected in the act-manner construction in the soccer domain. In addition, the results of the analysis of the co-variation of collexemes in two different slots of the same construction seem to suggest that such interactions are determined by frame-semantic knowledge.

Keywords: construction, Construction Grammar, frame, Fisher-Yates exact test, covarying collexeme analysis

## 1. Introduction

The issue of the investigation of interactions between words and grammatical structures has received considerable attention in recent years. Previous studies have concentrated on the preferences or restrictions associated with one particular slot in the construction. Stefanowitsch and Gries, for example, have analyzed the relationship between a particular construction and the lexical items occurring in a particular slot provided by it, both in relation to the language as a whole (Stefanowitsch and Gries, 2003) and in relation to semantically or functionally near-equivalent constructions (Gries and Stefanowitsch, 2004a). In this work they have demonstrated that such relationships rest on the degree of semantic compatibility between the meaning of the construction and that of the lexical item occurring in a given slot of it.

So far, however, far too little attention has been devoted to possible associations between two or more such slots. Previous work has been limited to the investigation of potential interactions between two words occurring in two different slots of the into-causative, English possessive constructions, and the way-construction

(Gries and Stefanowitsch, 2004b; Stefanowitsch and Gries, 2005). Their research revealed that there are restrictions imposed upon different slots of a construction and that these restrictions rest on three types of semantic coherence: one based on frame-semantic knowledge, one based on semantic prototypes, and one based on image schemas.

The aim of this paper is to investigate correlations between lexical items occurring in two different slots of the play-manner construction in the domain of soccer, and to show that such correlations are determined by frame-semantic knowledge.

This paper is structured as follows. In section 2, theoretical and methodological issues are addressed. In section 3, data, material, and tools are discussed. In section 4, a statistical procedure is described. Section 5 provides an account of the play-manner construction in the soccer domain. Section 6 presents and discusses the results. Section 7 gives an evaluation of the results and makes some proposals for future research.

## 2. Covarying collexeme analysis: theoretical and methodological frameworks

Covarying collexeme analysis is based on two frameworks, one theoretical and one methodological. The theoretical framework is provided by a constructional approach to grammar, in particular, the version of Construction Grammar developed by Goldberg (1995). From this perspective, *construction* is defined as any symbolic unit, complex or specific, that is to some degree arbitrary and unpredictable in respect of meaning or form. In this regard, Goldberg's Construction Grammar is different from other closely related linguistic approaches. In her later work, however, Goldberg (2006, p. 5) adopts a different standpoint such that she accepts compositionality on condition that the frequency of a pattern stored as a construction is large enough for it to be entrenched in a speaker/hearer's linguistic system. From this new viewpoint, constructions are identical to symbolic units in Cognitive Grammar (Langacker, 1987). Hence, constructions in this sense are viewed as symbolic units that combine form with meaning, where both form and meaning can differ in complexity and schematicity. The former covers morphology, lexis, syntax and even phonology and prosody, whereas the latter includes semantic, pragmatic and discourse-functional properties.

The methodological framework is offered by quantitative corpus linguistics. The corpus-based method is used to investigate correlations between lexical items occurring in two different slots of the play-manner construction: in other words, to identify which words are strongly attracted or repelled by particular slots in this construction. In contrast to traditional collocation-based techniques, covarying collexeme analysis allows us to determine the frequency of a word in one slot of a particular construction in relation to the words occurring in a different slot of the same construction. Although this approach is strictly quantitative and objective with respect to the way in which the strength of association between a particular construction and the words is identified, the results of such an analysis are evaluated qualitatively and subjectively. In particular, lexical items that are strongly attracted to a given (slot of a) construction can be grouped into semantic classes on the basis of introspective judgments.

## 3. Data, material, and tools

The primary source of data used in this analysis comes from specialized corpora collected in the Polish language, covering the years between 2005 and 2013. These corpora comprise different types of texts (such as official news, comments, biographies, written interviews with people linked with the sport, match reports and reviews, etc.) derived from various internet websites. The Polish corpus contains approximately 1 150 300 words. A software program, MonoConc Pro, was used to retrieve the observed frequencies from the corpus. The rest of the figures and expected frequencies were worked out by means of Microsoft Excel spreadsheets. All values required for the calculation of the association strengths were entered in the 2-by-2 table and submitted to the Fisher exact test. The p-value provided by this test was taken as an indicator of association strength, i.e., a word's strength of attraction/repulsion to a construction: the smaller the p-value, the stronger the association. This statistical analysis was carried out by means of an on-line Fisher's exact test calculator for two-by-two contingency tables.

## 4. Procedure

Let us exemplify this procedure using the Polish verb *grać* (*to play*) in the verb slot together with the manner adjunct (e.g. *słabo, poorly* in English) in the adjunct slot in the act-manner construction. Table 1 below shows the actual frequencies required for a covarying collexeme analysis of the verb *grać* and the manner adjunct *słabo* in the act-manner construction (for expository purposes, it also gives the expected frequencies for the verb *grać* and the manner adjunct *słabo* in parentheses). The figures in italics are derived from the corpus while the other figures are the results of additions and subtractions.

The first step of this procedure involves working out the observed frequencies. These are calculated in the following way. First, all occurrences of the act-manner construction are identified from the corpus: 2215. Second, the frequency of the verb *grać* in the verb slot is determined: 714. Third, the frequency of the adjunct *słabo* in the adjunct slot is extracted: 40. Finally, the frequency of the verb *grać* and the manner adjunct *słabo* is counted: 31. These four values are derived from the corpus directly while the remaining ones (that is, the frequency of all other verbs and the adjunct *słabo*: 9, the frequency of the verb *grać* and other manner adjuncts in the act-manner construction: 683, the frequency of all other verbs and other manner adjuncts in the act-manner construction: 1492, the total frequency of all other manner adjuncts in the act-manner construction: 2175, the total frequency of all other verbs in the act-manner: 1501) result from addition and subtraction.

*Table 1: The distribution of* grać *and* słabo *in the act-manner construction*

| | *Grać in verb slot of act-manner construction* | *All other verbs in verb slot of act-manner construction* | *Total* |
|---|---|---|---|
| *Słabo in adjunct slot of act-manner* | *31* (12.89) | 9 | *40* |
| *Other manner adjuncts in adjunct slot of act-manner* | 683 | 1492 | 2175 |
| *Total* | *714* | 1501 | *2215* |

These observed values now allow us to calculate the expected frequency of the verb *grać* and the manner adjunct *słabo* in the act-manner construction. The way in which this is done is actually quite simple. For the verb *grać* and the manner adjunct *słabo* in the act-manner construction, you simply have to multiply its column total by its row total, and divide the result by the overall table total. For example, for the top left cell in Table 1 – the one containing the figure 31, you take the column total (714) and multiply it by the row total (40). This gives you the rather large figure (28560). You then divide this figure by the table total (2215), giving you the result (12.89). In the case of my analyses, these calculations were performed in Microsoft Excel. If the observed frequency of the verb *grać* and the manner adjunct *słabo* in the act-manner construction is significantly higher or lower than expected, the relation between the verb *grać* and the manner adjunct *słabo* is one of attraction or repulsion respectively (the verb *grać* and the manner adjunct *słabo* are then considered to be significantly attracted or repelled *collexemes* of the act-manner construction).

The third step of this procedure involves working out the association strength of the verb *grać* and the manner adjunct *słabo*. In order to calculate the association strength between pairs of lexical items (the verb *grać* and the manner adjunct such as *słabo*) occurring in two different slots of the same construction, the following four frequencies need to be employed: the observed frequency of the verb *grać* and manner adjunct (e.g. *słabo*) in the act-manner construction; the frequency of all other verbs and manner adjunct (e.g. *słabo*) in this construction; the observed frequency of the verb *grać* and other manner adjuncts in the construction; the frequency of all other verbs and other manner adjuncts in the construction. These are entered in a two-by-two table and examined by means of the Fisher exact test. The p-value resulting from the calculation of Fisher exact test for this distribution is exceptionally small: 3.88E-09. This indicates that the verb *grać* and the manner adjunct *słabo* are highly significantly attracted to each other in the act-manner construction, but it can only be determined by comparing the observed frequencies of the verb *grać* and its manner adjunct with the expected ones. As this comparison shows, the verb *grać* occurs more frequently than expected with the manner adjunct *słabo* in the act-manner construction. In other words, *grać* and *słabo* are highly significant, very strongly attracted collexemes of the act-manner construction.

## 5. Act-manner construction

Let us now consider the Polish verb *grać* and the manner adjunct in the act-manner construction. This abstract construction can be represented structurally as [SUBJ [V ADJUNCT]], where V is a dynamic verb and ADJUNCT denotes manner. The manner adjunct is realized in this construction as an adverb, a noun phrase, or a prepositional phrase, as can be evident from the examples (a-c) below. The basic meaning of the construction is that an agent acts in a certain manner.

a) [*Wisła*] $_{\text{Subject, Team}}$ *grała* [*z kontry*] $_{\text{Adjunct, Manner.}}$ (Wisła played on the counter-attack).
b) [*Obydwie ekipy*] $_{\text{Subject, Team}}$ *grały* [*bardzo ofensywnie*] $_{\text{Adjunct, Manner.}}$ (Both teams were playing very offensively).
c) [*Reprezentant Niemiec*] $_{\text{Subject, Team}}$ *broni swojego zespołu, twierdząc, że we wtorek zagra* [*tak samo, jak w każdym innym spotkaniu*] $_{\text{Adjunct, Manner.}}$ (The German national player defends his team, claiming that the team will play on Tuesday the same as in each encounter).
d) [*Drużyny*] $_{\text{Subject, Team}}$ *grają* [*na najniższym poziomie*] $_{\text{Adjunct, Manner}}$ *w Europie.* (Teams play at the lowest level in Europe).

The semantics of the examples above is a little more specific than the afore-mentioned definition of the act-manner construction seems to suggest. That is, they are used in situations where a player or a team plays football in a certain manner. Manner elements belonging to the manner categories may evaluate how a player or a team plays, or describe in what way the action expressed in the VP is performed. If they evaluate the play, then lexical items evoke the PERFORMANCE EVALUATION frame. In this frame the player's or team's performance is judged by the speaker (the judge) against the performance of an opposing player or an opposing team. The player or the team who is involved in performing the action is assigned the same evaluation as the action itself. Thus, the player's or team's performance is considered to be of a certain quality (either good or bad). However, if the manner expressions describe in what way the action expressed in the VP is performed, then lexical items evoke the PERFORMANCE frame. In this frame the player or the team performs the act of playing football in a certain manner. This act is identified with the player's or team's performance. There are also some examples of manner elements, mentioned below, that come from the TACTICS frame, the QUALITY frame, the EMOTION DIRECTED frame, the DETERMINATION frame, the STATE OF NUMERICAL AMOUNT OF PLAYERS frame, the IDENTICALITY frame, and the SIMILARITY frame.

Given these semantic specifications, it is possible to predict roughly what verbs and manner adjuncts are likely to occur in both slots of this construction. The verbal slot should prefer verbs denoting the actions performed by an agent in a particular manner (in the case of this analysis, the Polish verb *grać*). The manner slot should prefer manner adjuncts belonging to different frames noted above. These predictions will be tested below.

What is important at this stage of the discussion, however, is that even the detailed description of the construction's semantics does not allow us to predict whether and how different slots in this construction are related semantically. It follows from the principle of semantic compatibility (Stefanowitsch and Gries, 2005, p. 11) that co-occurrences of the verb *grać* and manner adjuncts are expected to be semantically coherent, but it does not specify what kind of semantic coherence we could expect for this construction. It will be argued that the semantic coherence between two different slots of this construction is determined by frame-semantic knowledge.

The last issue that needs to be addressed is what syntactic forms manner elements may have. Prototypically, manner elements in English and Polish have the form of an adverb (e.g. to play *offensively*, grać *ofensywnie*). They may also have the form of NPs (e.g. to play *offside;* grać *pressingiem*) or, more often, PPs. Prepositions heading manner phrases include *with, in, without* and *through* in English (e.g. to play *with confidence*), as well as *w, na, bez* and *z* in Polish (e.g. grać *w tym samym stylu*). As far as manner PPs are concerned, they can be also characteristically realized by PPs headed by comparative *as* or *like* in English and *jak* in Polish. In addition, in a few cases we find NPs rather than PPs. This is usually found in examples in English in which the manner adjunct refers to tactics adopted by a team.

## 6. Results and discussion

Table 2 below shows the frequencies required to calculate the direction of association and the strength of association between pairs of lexical items occurring in two different slots of the act-manner construction. It also provides the expected frequency for verb *grać* and manner adjunct (e.g. *słabo*) in this construction: (a), as well as the results of a co-varying collexeme analysis ($P_{\text{Fisher exact}}$) for the 20 most strongly attracted combinations of the verb *grać* and manner adjuncts. The figures (a, e, x, z) were derived from a corpus directly, the other figures (b, c, d, f, y) are the results of additions and subtractions.

The results confirm the predictions made above, in that the semantic coherence between two different slots of the same construction is based on the relationship between two frames, i.e., on frame-semantic knowledge. Moreover, the specific suggestions concerning the meaning of the act-manner construction are also confirmed. For this construction, we find that the two most strongly associated pairs of covarying collexemes are *grać słabo* (*to play poorly*) and *grać na* [...] *poziomie* (*to play at a* [...] *level*) which instantiate the relationship between the PLAY frame and the QUALITY frame: in other words, they evoke the PERFORMANCE EVALUATION frame mentioned above. The p-values resulting from the calculation of Fisher exact for these collexeme pairs are exceptionally small: 3.88E-09 and 1.21E-08. If we take all significant covarying-collexeme pairs into consideration, it turns out that this relationship is in fact the predominant one in the construction. Seventh-ranked collexeme pair *grać lepiej* (*to play better*) and *grać z polotem* (*to play with panache*) also belong to this frame. Their indices of co-varying collostructional strength are 0.00051331 and 0.01073543. Comparing the observed and the expected frequencies of pairs of lexical items occurring in two different slots indicates that these collexeme pairs occur more frequently than expected in the act-manner construction. In other words, they are highly significant, very strongly attracted to each other in this construction. Note also that *grać słabo* is the most strongly associated covarying-collexeme pair in the construction, since its p-value is exceptionally small (3.88E-09), and the comparison of the observed and expected frequencies indicates that this collexeme pair occurs more frequently than expected in the act-manner construction.

*Table 2: The results of co-varying collexeme analysis for the act-manner construction*

| act-manner | a | x | e | z | b | c | y | f | d | (a) | $P_{\text{Fisher exact}}$ |
|---|---|---|---|---|---|---|---|---|---|---|---|
| grać słabo/ słabiej | 31 | 40 | 714 | 2215 | 9 | 683 | 2175 | 1501 | 1492 | 12.89 | 3.88E-09 |
| grać na (..) poziomie | 16 | 16 | 714 | 2215 | 0 | 698 | 2199 | 1501 | 1501 | 5.16 | 1.21E-08 |
| grać w osłabieniu | 16 | 16 | 714 | 2215 | 0 | 698 | 2199 | 1501 | 1501 | 5.16 | 1.21E-08 |
| grać na (…) możliwości | 18 | 19 | 714 | 2215 | 1 | 696 | 2196 | 1501 | 1500 | 6.12 | 1.62E-08 |
| grać w (liczba zawodników) | 20 | 23 | 714 | 2215 | 3 | 694 | 2192 | 1501 | 1498 | 7.41 | 7.40E-08 |
| grać ofensywnie | 22 | 29 | 714 | 2215 | 7 | 692 | 2186 | 1501 | 1494 | 9.35 | 1.54684E-06 |
| grać lepiej | 54 | 115 | 714 | 2215 | 61 | 660 | 2100 | 1501 | 1440 | 37.07 | 0.00051331 |
| grać tak, jak … | 13 | 18 | 714 | 2215 | 5 | 701 | 2197 | 1501 | 1496 | 5.80 | 0.000563818 |
| grać defensywnie | 9 | 11 | 714 | 2215 | 2 | 705 | 2204 | 1501 | 1499 | 3.55 | 0.001014914 |
| grać agresywnie | 9 | 12 | 714 | 2215 | 3 | 705 | 2203 | 1501 | 1498 | 3.87 | 0.002900886 |
| grać szybko/ szybciej | 11 | 16 | 714 | 2215 | 5 | 703 | 2199 | 1501 | 1496 | 5.16 | 0.002902424 |
| grać widowiskowo | 5 | 5 | 714 | 2215 | 0 | 709 | 2210 | 1501 | 1501 | 1.61 | 0.003447379 |
| grać tak samo | 7 | 9 | 714 | 2215 | 2 | 707 | 2206 | 1501 | 1499 | 2.90 | 0.006626995 |
| grać ambitnie | 8 | 11 | 714 | 2215 | 3 | 706 | 2204 | 1501 | 1498 | 3.55 | 0.006902564 |
| grać z polotem | 4 | 4 | 714 | 2215 | 0 | 710 | 2211 | 1501 | 1501 | 1.29 | 0.01073543 |
| grać zachowawczo | 4 | 4 | 714 | 2215 | 0 | 710 | 2211 | 1501 | 1501 | 1.29 | 0.01073543 |
| grać na (..) luzie | 4 | 4 | 714 | 2215 | 0 | 710 | 2211 | 1501 | 1501 | 1.29 | 0.01073543 |
| grać według.. | 5 | 6 | 714 | 2215 | 1 | 709 | 2209 | 1501 | 1500 | 1.93 | 0.015154429 |
| grać bojaźliwie | 3 | 3 | 714 | 2215 | 0 | 711 | 2212 | 1501 | 1501 | 0.97 | 0.033399116 |
| grać zespołowo | 3 | 3 | 714 | 2215 | 0 | 711 | 2212 | 1501 | 1501 | 0.97 | 0.033399116 |

**a** = Observed frequency of verb (*grać*) and manner adjunct (e.g. *słabo*) in the act-manner construction; **b** = Frequency of all other verbs and manner adjunct (e.g. *słabo*) in this construction; **c** = Observed frequency of verb (*grać*) and other manner adjuncts in the construction; **d** = Frequency of all other verbs and other manner adjuncts in the construction; **e** = Total frequency of verb (*grać*) in the construction; **f** = Total frequency of all other verbs in the construction; **x** = Total frequency of manner adjunct (*słabo*) in the construction; **y** = Total frequency of other manner adjuncts in the construction; **z** = Total frequency of the act- manner construction; **(a)** = Expected frequency of verb (*grać*) and manner adjunct (e.g. *słabo*) in the act-manner construction; $\mathbf{P}_{\text{Fisher exact}}$ = index of co- varying collostructional strength

*Grać słabo* and most other strongly associated covarying-collexeme pairs encode the act of playing football in a certain manner. All of them instantiate one of the following relationships between frames. For example, third-ranked *grać w osłabieniu* (*to play at a numerical disadvantage*) and fifth-ranked *grać w [liczba zawodników]* (*to play with [nine, ten, etc.] men*) instantiate the relation between the PLAY frame and the STATE OF NUMERICAL AMOUNT OF PLAYERS frame. Fourth-ranked *grać na [...] możliwości* (*to play as [...] as one can*) instantiates the relation between the PLAY frame and the CAPABILITY frame. *Grać ofensywnie* (*to play offensively*) and *grać defensywnie* (*to play defensively*) evoke the TACTICS frame in which a team uses defensive or offensive tactics. *Grać szybko* (*to play fast*) reflects cognitive knowledge concerning understanding of the speed at which the act of playing is performed, and hence it instantiates the relationship between the MOTION frame and the SPEED DESCRIPTION frame. *Grać tak, jak* (*to play as*) and *grać tak samo* (*to play the same as*) instantiate the relationship between the PLAY frame and the SIMILARITY frame, and between the PLAY frame and the IDENTICALITY frame respectively. *Grać na luzie* (*to play in a relaxed way*) and *grać bojaźliwie* (*to play timidly*) evoke the relations between the PLAY frame and the EMOTION DIRECTED frame. The verb *grać* and manner adjuncts instantiating these relations describe an experiencer (a team or a player) who is feeling or experiencing a particular emotional response to a stimulus (e.g. the play of opposing team, the state of a match). There can also be certain circumstances (e.g. bad play) under which the emotions occur or a reason that evokes these emotions. Finally, the collexeme pairs such as *grać agresywnie* (*to play aggressively*) and *grać ambitnie* (*to play with determination*) instantiate the relationship between the PLAY frame and the DETERMINATION frame.

The strong relation between the afore-mentioned frames clearly reflects cognitive and cultural knowledge about the way in which the act of playing is performed by a team or a player. This knowledge is based on both sensory-motor and cultural experience. The way of performing the act of playing may depend on the play of an opposing team, the score in the match, circumstances under which the act occurs, a reason why it occurs, and the purpose for which the player or the team performs the act of playing.

In the context of investigating the relationship between words occurring in two different slots of the act-manner construction, it may also be useful to examine collexeme pairs that are not significantly attracted to each other in this construction. The results of a co-varying collexeme analysis for the 5 most strongly repelled combinations of the verb *grać* and manner adjuncts are shown in Table 3. Amazingly, in spite of the same relationships between semantic frames they evoke, five pairs of lexical items such as *grać w* [...] *sposób* (*to play in a* [...] *way*), *grać bez* (*to play without*), *grać pewnie* (*to play confidently*), *grać fatalnie* (*to play miserably*), and *grać spokojnie* (*to play calmly*) are strongly repelled collexeme pairs, since their p-values resulting from the calculation of Fisher exact are very high: 1; 1; 1; 0,999999539; 1,00E+00. Furthermore, comparing the observed and the expected frequencies for each pair of words confirms that these collexeme pairs occur less frequently than expected in the construction. Thus, they are strongly repelled pairs of covarying (verb + manner adjunct) collexemes.

*Table 3: The results of co-varying collexeme analysis for the 5 most strongly repelled combinations of the verb* grać *and manner adjuncts*

| | a | x | e | z | b | c | y | f | d | (a) | $P_{Fisher\ exact}$ |
|---|---|---|---|---|---|---|---|---|---|---|---|
| grać spokojnie | 6 | 58 | 714 | 2215 | 52 | 708 | 2157 | 1501 | 1449 | 18,70 | 1,00E+00 |
| grać fatalnie | 2 | 45 | 714 | 2215 | 43 | 712 | 2170 | 1501 | 1458 | 14,51 | 0,999999539 |
| grać pewnie | 3 | 96 | 714 | 2215 | 93 | 711 | 2119 | 1501 | 1408 | 30,95 | 1 |
| grać bez... | 15 | 176 | 714 | 2215 | 161 | 699 | 2039 | 1501 | 1340 | 56,73 | 1 |
| grać w (..) sposób | 3 | 123 | 714 | 2215 | 120 | 711 | 2092 | 1501 | 1381 | 39,65 | 1 |

## 7. Conclusion

In conclusion, the results of this investigation reveal that the semantic coherence between the most strongly associated covarying collexeme pairs in the act-manner construction is based on semantic frames. The covarying collexeme analysis has shown not only the high degree of semantic coherence that exists between two different slots of the same construction, but also systematic relationships between semantic frames that determine this semantic coherence. These relationships are clearly not the exception, but the rule for the act-manner construction in the domain of soccer. This analysis also has indicated that there are strongly attracted and repelled pairs of covarying collexemes of this construction.

The method applied in this analysis can have a wide range of applications in language study and language pedagogy. First, it can be employed for developing linguistic theory. Second, it can be used to increase the precision of language account by providing more precise results than traditional collocate-based-approaches. Finally, it can be adopted for the purpose of identifying the meaning of a grammatical construction and determining which lexical items or pairs of lexical items are strongly associated with or repelled by a particular construction. To adopt this approach further, future research will have to provide a complete analysis of constructions and all lexical units occurring in their different slots, and a detailed account of all frames and conceptual metaphors evoked by those constructions and words. The possible use of this method is much wider, and future research will show the full extent of this potential.

## References

*Goldberg, A.* (1995). *Constructions: A construction grammar approach to argument structure.* Chicago: Chicago University Press.

*Goldberg, A.* (2006). *Constructions at work. The nature of generalization in language.* Oxford: Oxford University Press.

*Gries, S. Th. and A. Stefanowitsch.* (2004a). Extending collostructional analysis: A corpus-based perspective on alternations. *International Journal of Corpus Linguistics,* 9(1), 97–129.

*Gries, S. Th. and A. Stefanowitsch.* (2004b). Co-varying collexemes in the into causative. In M. Achard and S. Kemmer (Eds.), *Language, culture, and mind* (pp. 225–236). Stanford, CA: CSLI.

*Langacker, R.* (1987). *Foundations of cognitive grammar.* Vol. 1. *Theoretical Prerequisites.* Stanford, CA: Stanford University Press.

*Stefanowitsch, A. and S. Th. Gries.* (2003). Collostructions: Investigating the interaction between words and constructions. *International Journal of Corpus Linguistics,* 8 (2), 209–243.

*Stefanowitsch, A. and S. Th. Gries.* (2005). Covarying collexemes. *Corpus Linguistics and Linguistic Theory,* 1 (1), 1–43.

Jacek Tadeusz Waliński
*University of Lodz, Poland*[1]

# Time in structuring fictive motion: an empirical corpus-based study

**Abstract:** This study demonstrates a role of time in structuring fictive motion expressions on the basis of empirical language data found in the National Corpus of Polish (NKJP). It focuses on a specific category of fictive motion used to depict forms, orientations, or locations of spatially extended objects in terms of a path over the object's extent. Although fictive motion employs event structuring, frequencies of language patterns found in the corpus demonstrate that language users tend to avoid articulating such expressions in temporal terms. The avoidance of temporality in fictive motion can be attributed to a basic conceptual difference between objects and force-dynamic events, as distinguished in Langacker's canonical event model. According to that model, while time is the primary domain of conceptualization for events, for objects the primary domain of conceptualization is space.

Keywords: fictive motion, coextension paths, events, time, space

## 1. Fictive motion

*Fictive motion* (Talmy, 1996, 2000; Langacker, 2005, 2008a) refers to figurative representations of motion attributed to immobile material objects, states, or abstract concepts, in which the meaning of motion verbs is semantically extended to express relations that do not involve motion *per se*, nor change of state. The following sentences (1) and (2) illustrate a specific category of fictive motion used for representation of static objects in space, which was labeled by Talmy (2000) as *coextension paths*. A coextension path refers to "a depiction of the form, orientation, or location of a spatially extended object in terms of a path over the object's extent" (Talmy, 2000, p. 138).

(1) The road goes downhill steeply into the valley.
(2) This wire fence goes all the way down to the wall.

1 Research supported with grant no. 2011/01/M/HS2/03042 from Polish National Science Centre.

As illustrated by the above examples, the described object is stationary and there is no entity traversing the depicted path, however, it is represented as moving along or over its spatial configuration.

Talmy (2000) notes that explanation of fictive motion in terms of metonymy would be inadequate because numerous cases of non-travelable paths cannot be associated with motion, e.g. *fences* do not move. Talmy (2000, pp. 100–104) considers fictive motion as *non-veridical*. He attributes the discrepancy between static and dynamic interpretations of fictive motion expressions to a distinction between *fictive* and *factive* modes of cognition. The former is more perceptually salient but less veridical, while the latter is more veridical but less perceptually salient. Thus, the "ception"—a neologism coined by Talmy to refer both to perception and conception—of the fictive mode requires perceptual veridicality to be overridden, which occurs naturally due to a general preference for dynamism in linguistic, perceptual, and conceptual semantics.

Matsumoto (1996) points out some intriguing characteristics of fictive motion expressions from the perspective of a cross-linguistic comparison between English and Japanese. He makes a distinction between *travelable paths*, i.e. paths that can be traveled by people, e.g. roads, paths, etc., as in (1), and *non-travelable paths*, i.e. paths including objects that do not normally act as media of human motion, e.g. wires, fences, etc., as in (2). Matsumoto (1996, pp. 213–217) reports that while English expresses both these types, in Japanese some non-travelable entities, such as *walls* and *fences*, cannot be described with fictive motion. Some other non-travelable entities, such as *borders* and *wires*, take a restricted set of motion verbs. This can be motivated by the fact that in Japanese certain motion verbs cannot be used to describe movement of a path that does not involve a perceptuo-motor basis. Rojo and Valenzuela (2009, Exp. 1) do not observe that distinction to occur as vividly in Spanish, but detect that it takes longer for Spanish speakers to process fictive motion sentences with non-travelable entities than parallel sentences with travelable ones.

Moreover, Matsumoto (1996, p. 204) distinguishes two types of fictive motion expressions. Type I includes sentences in which the motion is arbitrary in the sense that it does not occur at any specific time. Type II is associated with an actual experience of motion of the person uttering the sentence (cf. *global vs. local view* in Langacker, 2005, p. 176; 2008b, pp. 69–70). Matsumoto (1996, p. 205) adds that "perspective mode and scope of attention are not necessarily correlated with the distinction between the motion of a particular entity at a particular time and the motion of an arbitrary entity that can be evoked at any time".

Langacker (2005, 2008a, 2008b) assumes that fictive motion reflects subjective imaginative mental constructions used to discuss actual existence of objects in real-life scenarios. It involves *mental scanning*, i.e. a partial reactivation of the original experience conceptualized along the imagined trajectory. Langacker (2008a, p. 529) argues that both expressions of actual and fictive motion involve scanning along a path. In actual motion we conceptualize events by tracking a mover's progress along a spatial path that it traverses physically: in *processing time* the mover is successively conceptualized as occupying a series of locations that collectively constitute the path. Accordingly, a motion event is apprehended by the conceptualizer by mentally accessing the successive locations in the same order that the mover reaches them. The movement occurs in *conceived time*, which is inherently involved in any event (cf. Radvansky & Zacks, 2011).

Essentially, in fictive motion the same mental operations are applied to a static scene, where a conceptualization occurs in a manner generally parallel to conceptualizations of actual motion events. An analog of the mover is a spatially extended object. Instead of tracking an object's movement, the conceptualizer scans mentally along the object's path, by which they invoke the constitutive locations to build up to a full conception of the object's spatial configuration. The *conceived time* has no significant role in the expression's objective content, because the object occupies all spatial locations on the path simultaneously. Langacker (2008a, pp. 83, 111–112) proposes to term this mode of building up gestalts manipulabe as simultaneously available wholes as *summary scanning*.

However, Jackendoff (2002, p. 362) questions whether any sort of scanning is involved in processing fictive motion sentences. He notes that although the sense of an observer scanning the extended object has some intuitive appeal, it does not account for the difference in inference patterns used in formal decomposition of state-functions. Jackendoff (1983, p. 173; 2002, pp. 360–362) argues that the conceptualization of fictive motion is associated with static representations of *Paths*, which designate directions, shapes, orientations, etc. Paths themselves are atemporal and can appear as arguments of state-functions. From this perspective, coextension path expressions should be viewed as instances of the state variant of GO-function, namely *non-temporal extension*, in which all points of the object's spatial configuration are activated simultaneously. Jackendoff (1983, 2002) suggests two possibilities concerning the relationship between GO and $GO_{EXT}$. Either they are not distinct functions and the difference depends on whether the GO-function is a feature of an [EVENT] or a [STATE], or alternatively they are distinct functions, but share a common internal structure. Iwata's (1996) detailed analysis of motion/extent as two semantic variants of the GO-function provides support for the latter option.

## 2. Temporality of fictive motion

Matlock (2004a, p. 1390) argues that fictive motion sentences frequently incorporate words and phrases communicating physical movement, which can be denoted in terms of duration, e.g. "The road runs along the coast *for 10 minutes*". Additionally, the natural temporality of physical movement is implied by those fictive motion expressions that include verbs communicating fast or slow manners of motion, e.g. "The freeway *races* past the city" or "Interstate 5 *crawls* through Los Angeles". The assumption of temporality of fictive motion expressions is elaborated further in Matlock's (2004b) discussion on a conceptual motivation of fictive motion constructions. She argues that objects depicted with fictive motion must be sufficiently long to dynamically construe over time for the mental scanning to occur. Matlock (2004b, p. 229) argues that sentences such as (3a) are much more natural than such as (3b) because of the temporality involved in the scanning along the path of a substantially longer object.

(3) a. The road runs along the coast *for 2 hours*.
b. ? The road runs along the coast *for 2 seconds*.

She adds, however, that sentences such as (3b) are reasonably plausible, if the intention of the speaker is to contrast a particularly short part of the road in question with other sections of the same road (see also Gibbs & Matlock, 2008 for a discussion of fictive motion from the perspective of *embodied simulation*).

Matsumoto (1996, p. 186–187) points out that in sentences such as "The highway runs along the coast for a while" the temporal phrase *for a while* denotes the duration of the process of motion along the coast, which is directly correlated to the length of the relevant section of the highway in question. He emphasizes that in such sentences temporality cannot be attributed to any particular duration of the state of location, which indicates that conceptualizations of fictive motion cannot be explained without assuming some sort of motion processing (see also Zwaan, 2009).

However, Blomberg and Zlatev (2013) argue that neither cognitive linguistic models proposed by Talmy (1996, 2000) and Langacker (2005, 2008a), nor account for fictive motion in terms of mental simulation proposed by Matlock (2004a, 2004b) adequately explain the experiential and linguistic complexity of the phenomenon. They point out that from a phenomenological perspective the view of fictive motion as grounded in mental simulation of movement does not make clear what is actually simulated because different conceptualizations of fictive motion may be motivated by at least three different features of human consciousness: *enactive perception*, *visual scanning*, and *imagination*. Moreover, they emphasize that fictive motion structuring may be related to *sedimentation of meaning*, understood as the linguistic consolidation of cognitive structures originally given in embodied sense-experience through certain persisting linguistic conceptualizations superimposed by language acquisition and socio-cultural transmission (Woelert, 2011). Because explanation of fictive motion in terms of mental simulation obscures these aspects, Blomberg and Zlatev (2013) postulate that the full account for fictive motion should be based on a broader phenomenological-linguistic framework, which takes into account *consciousness-language interactionism* as a complex set of reciprocal relations between pre-linguistic experience and linguistic meaning.

## 3. A corpus-based study of temporality in structuring fictive motion

This study aims to observe how the assumption of temporality in fictive motion expressions fits into the reality of linguistic performance. More specifically, it is based on an assumption that if the *conceived time* plays a significant role in cognitive processing of coextension paths, such expressions should feature a significant number of temporal representations in language corpora, which has been observed for actual motion events (cf. Waliński, 2013). This paper parallels a study conducted for English with the BNC (Waliński, 2014a) and systematically expands preliminary research conduced earlier (Waliński, 2013, Ch. 8) by employing a six times as extensive corpus and doubling the number of lexical combinations taken into consideration.

This paper is positioned within a *cognitive corpus-based* approach to language study, which brings together the descriptive framework of *cognitive linguistics* (Croft & Cruse, 2004; Evans, 2012; Kardela, 2006) with the methodological workbench of *corpus linguistics* (McEnery & Hardie, 2012). Essentially, it relies on explanatory notions adopted by the cognitive linguistics framework, but approaches them in such a way that their relevance to a given linguistic phenomenon can be empirically validated in large corpora (Heylen, Tummers & Geeraerts, 2008; see Lewandowska-Tomaszczyk & Dziwirek, 2009; Glynn & Fischer, 2010 for edited collections of studies). This study employs the full archive of *Narodowy Korpus Języka Polskiego* (the National Corpus of Polish, henceforth, the NCP), which is a 1.5 billion word collection of samples of both spoken and written contemporary Polish (Przepiórkowski, et al., 2012; see www.nkjp.pl for more information). The

NCP has an important advantage of being a publicly available *standard reference* corpus (McEnery & Wilson, 2001, p. 32), which enables other researchers to attest or expand the present research with nothing else than a web browser.

Fictive motion expressions are problematic to pick out from language corpora because at the syntactic level they are virtually indistinguishable from actual motion expressions. For that reason, the investigation was implemented with a procedure that involves looking for combinations of landmarks that can potentially feature in coextension paths with an ample selection of motion verbs.

Selecting suitable landmarks followed observations (Langacker, 2005; Matlock, 2004b) that coextension paths typically describe extended or elongated stationary spatial entities. Starting with a few prototypical ones, such as road, wire, fence, coast, etc. (PL: *droga, przewód, płot, plaża*), the online version of *SłowoSieć* 2.1 – an interface to Polish wordnet (Piasecki, et al., 2009; see plwordnet.pwr.wroc.pl for more information) was consulted to identify other spatially extended objects potentially fit for descriptions with coextension paths. The following four basic categories of landmarks were selected for analysis:

a) Travelable paths, such as footpath, highway, road, bridge, tunnel, viaduct, way, etc.: *aleja, alejka, arteria, autostrada, bulwar, chodnik, deptak, droga, dróżka, dukt, estakada, jezdnia, kładka, korytarz, most, mostek, obwodnica, pas, pasaż, pobocze, promenada, przejazd, przejście, przeprawa, szlak, szosa, ścieżka, tor, trakt, trasa, trotuar, tunel, ulica, uliczka, wiadukt*. These spatial entities are distinguished by Matsumoto (1996) as paths intended for traveling by people.
b) Travelable natural landmarks, such as beach, coast, desert, field, forest, mountain, valley, etc.: *bezdroża, brzeg, cieśnina, cypel, dolina, góra, grań, grzbiet, jar, kanał, kanion, klif, kotlina, las, lodowiec, łąka, masyw, moczary, nabrzeże, niecka, pagórek, parów, pasmo, perć, plaża, płaskowyż, pole, połonina, półwysep, przełęcz, przesmyk, przylądek, pustynia, puszcza, turnia, wądół, wąwóz, wertepy, wybrzeże, wysepka, wyspa, wzgórze, wzniesienie, zagajnik, żleb*. These typically extended or elongated landmarks can also be traveled, however they were not built intentionally for this purpose.
c) Non-travelable connectors, such as cable, line, pipe, pipeline, wire, etc.: *drut, instalacja, kabel, linia, łącze, łącznik, magistrala, przewód, rura, rurociąg*. These elongated objects are classified by Matsumoto (1996) as *non-travelable paths*, since they are not normally traveled by people. They include predominantly objects used for transmitting energy or transporting substances over long distance.
d) Non-travelable barriers, such as border, dam, fence, hedge, wall, etc.: *bariera, granica, miedza, mur, murek, obwałowanie, ogrodzenie, parkan, płot, siatka,*

*ściana, tama, wał, zapora, żywopłot*. These spatially extended entities are not normally used for traveling since they serve as barriers, but they often stretch over a relatively substantial distance.

Altogether 105 landmarks were selected. It is noteworthy that this set incorporates diminutive nominal forms commonly used by Polish speakers to refer to less conspicuous paths, e.g. *alejka, dróżka, uliczka, mostek, murek* [Lit. short/petite: avenue, path, street, bridge, wall] and includes lexicalizations of certain landmarks marked in English with modifiers, e.g. *ogrodzenie, płot, siatka* [fence, wooden fence, wire fence]. Since the range of objects that can potentially be described with coextension paths is practically unrestricted, the above selection is far from being exhaustive. Still, it seems to be reasonably adequate for the purpose of this study.

Suitable verbs of motion were selected with reference to a classification proposed by Levin (1993, Ch. 51). They were identified with the help of *SłowoSieć* 2.1 (2014) and the *PWN-Oxford Dictionary* (2004). The following categories of motion verbs were selected for analysis:

a) Verbs of inherently directed motion, e.g. advance, ascend, climb, come, cross, descend, fall, leave, plunge, return, rise, etc. Levin (1993: Ch. 51.1) notes that meaning of these verbs specifies the direction of motion, even if an overt directional complement is not present. The following verbs were selected: *dobiegać, dobrnąć, dochodzić, doczołgać, doczłapać, dolatywać, dopełznąć, dopłynąć, doskakiwać, dowlec, krzyżować, obniżać, odbiegać, odbijać, odchodzić, oddalać, odlatywać, odpłynąć, odskakiwać, omijać, opadać, opuszczać, pobłądzić, podchodzić, podkradać, podążać, pognać, poszybować, powlec, powłóczyć, przebiegać, przebrnąć, przechodzić, przecinać, przedzierać, przelatywać, przemierzać, przemykać, przeskakiwać, przesuwać, przetaczać, przybliżać, schodzić, spadać, uciekać, umykać, unosić, wchodzić, wdzierać, wjeżdżać, wracać, wsuwać, wybiegać, wychodzić, wyczołgać, wygramolić, wylatywać, wymijać, wymykać, wypełzać, wypływać, wyskakiwać, wysuwać, wzbijać, wznosić, wślizgiwać, zabrnąć, zabłądzić, zabłąkać, zachodzić, zagłębiać, zakradać, zapuszczać, zawlec, zawracać, ześlizgiwać, zjeżdżać, zmierzać, zsuwać.*
b) Verbs of motion manner do not indicate the direction of motion as an inherent part of their meaning (cf. Matsumoto, 1996 for restrictions on manner semantics in fictive motion). Instead, they express different manners or means of motion, which can basically be subdivided into two categories:

   i. Run verbs, e.g. climb, crawl, fly, glide, hike, journey, leap, meander, prowl, race, run, shuffle, stray, swim, travel, wade, wander, zigzag, etc.: *biegać, błądzić, błąkać, brnąć, chodzić, cwałować, człapać, czmychać, czołgać, dążyć, defilować, dreptać, galopować, gnać, gramolić, halsować, hasać, iść, koziołkować, kuleć, kuśtykać, lecieć, łazić, maszerować, mijać, mknąć, mozolić, paradować, pełzać, pędzić, płynąć, podróżować, potykać, rzucać, skakać, skradać, słaniać, smyrgać, spacerować, stąpać, sunąć, szarżować, szybować, ślizgać, śmigać, śpieszyć, truchtać, utykać, wałęsać, wędrować, wlec, włóczyć, wspinać, zbaczać, zmierzać.* Levin (1993, Ch. 51.3.2) notes that most of these verbs describe different motion manners of both animate and inanimate entities. She adds that this category probably requires a further subdivision.

ii. Roll verbs, e.g. roll, slide, swing, coil, revolve, rotate, spin, turn, twist, whirl, wind, etc: *huśtać, kołysać, kręcić, obchodzić, obiegać, obracać, okręcać, otaczać, owijać, skręcać, toczyć, turlać, wić, wirować, zakręcać, zataczać, zawijać.* Levin (1993: Ch. 51.3.1) notes that many verbs that describe motion around an axis take a restricted range of prepositions describing the path of motion.

c) Chase verbs, e.g. chase, follow, pursue, tail, track, etc. They are typically transitive, with the chaser as subject and the entity of pursuit as object (Levin, 1993, Ch. 51.6). The following verbs were selected: *gonić, ścigać, śledzić, tropić.*
d) Accompany verbs, e.g. accompany, conduct, escort, guide, lead, etc.: *ciągnąć, kierować, prowadzić, towarzyszyć, wieść.* These verbs basically relate to one person taking another from a place to place. Their meaning is differentiated by the nature of the relation between the two participants (Levin, 1993, Ch. 51.7).

Altogether 160 motion verbs were selected for analysis. The Polish set is somewhat less diverse than the selection of English verbs presented by Levin (1993, Ch. 51). As observed by Kopecka (2010: 241), "Polish does not exploit the slot of the main verb as productively as English does. In Polish, the size of the Manner verb lexicon, although still substantial, appears to be smaller, and the sorts of fine-grained semantic components of Manner lexicalized in the verbs are less diverse". It is also noteworthy that in Polish some verbs of inherently directed motion are closely related to the verbs of motion manner, as they incorporate perfectivizing prefixes, which developed historically from adnominal elements (cf. Tabakowska, 2003). It must be emphasized that correspondences between English and Polish verbs of motion appear to be rather convoluted and deserve a discussion that far exceeds the scope of this study.

Some categories of motion verbs distinguished by Levin (1993) are not included in the analysis. For example, *waltz verbs*, i.e. ones that are zero-related to names of dances and mean roughly "to perform the dance" (Levin, 1993, Ch. 51.5), e.g. *boogie, polka, rumba, tango,* etc., were excluded as too idiosyncratic. The use of *instrumental motion verbs* (Levin, 1993, Ch. 51.4) in coextension path expressions is normally restricted in fictive motion, which is discussed separately

(Waliński, 2014c). Although the above selection is far from being exhaustive, it seems to be reasonably adequate for the purpose of investigating the role of time in structuring fictive motion.

Essentially, the above sets allow for checking 16,800 different *landmark + motion verb* combinations (105 landmarks × 160 verb forms). Additionally, because the NCP offers an underlying morphological dictionary, which allows for querying all declensional forms for nouns and conjugational forms for verbs, the number of analyzed combinations is vastly greater. Since this study aims specifically to identify examples of coextension paths denoted in terms of duration, e.g. "The road runs along the coast for 10 minutes" vis-à-vis ones denoted in terms of distance, e.g. "The road runs along the coast for 10 kilometers", a unit of space/ time measurement was also incorporated in the following lexical pattern used for investigation:

**LANDMARK + MOTION VERB + TEMPORAL / SPATIAL UNIT; SLOP=3, PRESERVE ORDER=YES**

Units of time measurement selected for analysis involve those that are typically used to express duration, i.e. *seconds, minutes, hours,* and *days* [moment, sekunda, minuta, godzina, dzień] together with their commonly used abbreviations. Units of space measurement selected for comparison include *meter, kilometer* and *mile* [*metr, kilometr, mila*] together with their abbreviations.

Because lexemes from the above linguistic pattern do not always follow directly one after another in fictive motion expressions, searching was implemented with *proximity queries* (Bernard & Griffin, 2009). They allow for searches with a *slop* factor, which specifies how far apart lexical items included in a query can be from one another to be still returned as a result to the query. The slop factor can be used in combination with a binary (yes/no) *preserve order* option, which indicates whether the original order of query terms should be retained in results. In this study the search was implemented in a relaxed manner using the slop value of 3, but the *preserve order* option set to "yes" to prevent an excess of coincidental hits. A listing of all queries used to obtain the results together with corresponding concordances retrieved from the NCP has been published online as a research report (Waliński, 2014b), which provides for immediate replicability of the study.

## 4. Summary of results

Corpus queries based on the above pattern returned 1698 matching concordance lines from the NCP. The resulting set was reviewed to exclude coincidental matches, since the search was implemented using proximity queries with a large value of slop factor, which while increasing the recall, substantially decreases the precision of results (see Pęzik, 2011). In result, 351 sentences were recognized as fictive motion expressions denoting extension of spatial entities in either spatial or temporal terms. The results found for the selected categories of landmarks are presented in Table 1.

*Table 1: Fictive motion denoted in either spatial or temporal terms in the NCP*

| *Category of verbs* | **Coextension paths denoted in spatial terms** | **Coextension paths denoted in temporal terms** |
|---|---|---|
| Travelable paths | 114 | 1 |
| Travelable natural landmarks | 124 | 0 |
| Non-travelable connectors | 31 | 0 |
| Non-travelable barriers | 81 | 0 |
| **Total** | **350** | **1** |

As shown in Table 1, with one exception all expressions of fictive motion identified in the NCP are denoted in spatial terms, which was found for each selected category of landmarks. The only example of coextension path denoted in temporal terms (4) indicates that the use of temporality in fictive motion tends to be rather occasional in linguistic performance of Polish speakers.

(4) Ścieżka prowadzi teraz 15 min. północną stroną Iwanowskiego Potoku.
[EN: The path leads now for 15 minutes along the northern side of Iwanowski brook.]

Curiously enough, in that particular example temporality is additionally accentuated by the temporal marker *now*, which indicates that it might be more closely associated with Type II of fictive motion, as distinguished by Matsumoto (1996). However, the sole example found in the corpus is not sufficient to support any systematic claims.

Although it is impossible to discuss the living organism of language in absolute numbers, the results suggest that on the basis of linguistic intuition we cannot assume that fictive motion expressions frequently incorporate phrasing that communicates physical movement in terms of duration (cf. Matsumoto, 1996; Matlock, 2004a; 2004b). The absence of temporal representations in fictive motion expressions identified with the above discussed procedure in the National Corpus of Polish is particularly conspicuous when compared to a substantial proportion of temporal representations found for the *motion-framed distance* (cf. Waliński, 2013).

## 5. Conclusions

Results obtained in this study demonstrate a peripheral role of the *conceived time* in structuring fictive motion. Obviously, the *processing time* is still involved, since even atemporal conceptions of static relations in space are conceptualized through mental processing, which requires time to occur in the first place (cf. Langacker, 2005; 2008a, p. 79, 529). The findings confirm the *non-veridical* nature of fictive motion (Talmy, 1996; 2000, Ch. 2), which, despite involvement of practically identical verbs and syntactic structures as actual motion expressions, is used to describe relations that do not involve motion per se nor change of state. Although the surface linguistic structure of fictive motion expressions reflects a cognitive bias towards *dynamism*, the apparent atemporality of coextension paths in linguistic performance indicates that at the conceptual level they are often processed as figurative representations of non-temporal states of spatial extension (Jackendoff, 1983, p. 173; 2002, pp. 360–362; Iwata, 1996).

The apparent absence of temporality in coextension paths can be attributed to the basic conceptual difference between static physical objects and force-dynamic interactions conceptualized as events (Langacker, 2008, 2012; see also Engberg-Pedersen, 1999). Because all events occur dynamically along the temporal axis as well as along the spatial axis (Radvansky & Zaacks, 2011), it is natural for temporal representations to frequently figure in linguistic expressions of distance involving actual motion. For the same reason, because fictive motion events are non-veridical, it is not overly surprising that temporality does not feature frequently in coextension paths. Because space is fundamentally static and globally accessible in nature (Galton, 2011; Langacker, 2012), figurative descriptions of objects with fictive motion hardly ever require temporal representations.

The relative absence of temporality in fictive motion found in empirical linguistic data suggests that there exist two cognitive modes of processing fictive motion, which fits into the *overall framework of fictivity* proposed by Talmy (2000, pp. 99–103). The results obtained in psycholinguistic experiments seem to demonstrate the *fictive mode*, in which processing of fictive motion takes place in a manner to some extent parallel to actual motion. This mode has arguably a greater potential for denoting spatial extension in terms of duration, since it involves an association with physical movement. On the other hand, the atemporality of coextension paths found in language corpora points at the *factive mode*, in which coextension paths are processed as atemporal expressions of spatial extension.

The link between linguistic structuring and evocation of an actual sense or conceptualization of motion is difficult to draw precisely. As emphasized by Talmy (2000, pp. 104–105), for the same instance of fictive motion expression some speakers will report a strong semantic evocation of motion, while others will report that there is none at all. The choice of either factive or fictive mode of conceptualization may be related to the dynamic potential of the linguistic structure, which may far exceed the scope of a single sentence. Coextension path expressions used in depictions of spatial scenes, which are inherently dynamic and multidimensional, have a greater potential to engage the fictive mode than a single sentence used every now and then to express a simple relation of distance. The factive mode of processing can also be attributed to *sedimentation of meaning* (Woelert, 2011; Blomberg & Zlatev, 2013). Since fictive motion is a convenient means of denoting spatial relations readily available in the conceptual repertoire of language users, it is likely to be used to describe spatial configurations of objects, at least in a good number of cases, in a parrot fashion dissociated from movement.

## References

*Bernard, E., & Griffin, J.* (2009). Understanding Lucene's query syntax. In *Hibernate Search in Action* (pp. 202–214). Greenwich, CT: Manning.

*Blomberg, J., & Zlatev, J.* (2013). Actual and non-actual motion: why experientialist semantics needs phenomenology (and vice versa). *Phenomenology and the Cognitive Sciences*, 1–24. doi:10.1007/s11097-013-9299-x.

*Croft, W., & Cruse, D. A.* (2004). *Cognitive Linguistics*. Cambridge: Cambridge University Press.

*Engberg-Pedersen, E.* (1999). Space and Time. In J. Allwood & P. Gärdenfors (Eds.), *Cognitive Semantics: Meaning and Cognition* (pp. 131–152). Amsterdam: John Benjamins.

*Evans, V.* (2012). Cognitive linguistics. *Wiley Interdisciplinary Reviews: Cognitive Science*, *3*(2), 129–141. doi:10.1002/wcs.1163

*Galton, A.* (2011). Time flies but space does not: Limits to the spatialisation of time. *Journal of Pragmatics*, *43*(3), 695–703. doi:10.1016/j.pragma.2010.07.002

*Gibbs, R. W., & Matlock, T.* (2008). Metaphor, imagination, and simulation. In R. W. Gibbs (Ed.), *The Cambridge Handbook of Metaphor and Thought* (pp. 161–176). Cambridge: Cambridge University Press.

*Glynn, D., & Fischer, K. (Eds.).* (2010). *Quantitative Methods in Cognitive Semantics: Corpus-Driven Approaches*. Berlin: De Gruyter Mouton.

*Heylen, K., Tummers, J., & Geeraerts, D.* (2008). Methodological issues in corpus-based Cognitive Linguistics. In G. Kristiansen & R. Dirven (Eds.), *Cognitive*

*Sociolinguistics Language Variation, Cultural Models, Social Systems* (pp. 91–128). Berlin: Mouton de Gruyter.

*Iwata, S.* (1996). Motion and Extent: Two Sides of the Same Coin. *Studia Linguistica, 50*(3), 256–282. doi:10.1111/j.1467-9582.1996.tb00351.x

*Jackendoff, R.* (1983). *Semantics and Cognition*. Cambridge, MA: MIT Press.

*Jackendoff, R.* (2002). *Foundations of Language: Brain, Meaning, Grammar, Evolution*. Oxford: Oxford University Press.

*Kardela, H.* (2006). Metodologia językoznawstwa kognitywnego. In P. Stalmaszczyk (Ed.), *Metodologie językoznawstwa współczesnego: Podstawy teoretyczne* (pp. 196–233). Łódź: Wydawnictwo Uniwersytetu Łódzkiego.

*Langacker, R. W.* (2005). Dynamicity, fictivity, and scanning: The imaginative basis of logic and linguistic meaning. In D. Pecher & R. A. Zwaan (Eds.), *Grounding Cognition: The Role of Perception and Action in Memory, Language, and Thinking* (pp. 164–197). Cambridge: Cambridge University Press.

*Langacker, R. W. (*2008a). *Cognitive Grammar: A Basic Introduction*. Oxford: Oxford University Press.

*Langacker, R. W. (*2008b). Cognitive Grammar as a basis for language instruction. In P. Robinson & N. C. Ellis (Eds.), *Handbook of Cognitive Linguistics and Second Language Acquisition* (pp. 66–88). New York: Routledge.

*Langacker, R. W.* (2012). Linguistic manifestations of the space-time (dis)analogy. In L. Filipović & K. M. Jaszczolt (Eds.), *Space and Time in Languages and Cultures: Language, culture, and cognition* (pp. 191–216). Amsterdam: John Benjamins.

*Levin, B.* (1993). *English Verb Classes and Alternations: A Preliminary Investigation*. Chicago: University of Chicago Press.

*Lewandowska-Tomaszczyk, B., & Dziwirek, K. (Eds.).* (2009). *Studies in Cognitive Corpus Linguistics*. Frankfurt am Main: Peter Lang.

*Matlock, T. (*2004a). Fictive motion as cognitive simulation. *Memory & Cognition, 32*(8), 1389–1400. doi:10.3758/BF03206329

*Matlock, T. (*2004b). The conceptual motivation of fictive motion. In G. Radden & K.-U. Panther (Eds.), *Studies in Linguistic Motivation* (pp. 221–248). Berlin: Mouton de Gruyter.

*Matsumoto, Y.* (1996). Subjective motion and English and Japanese verbs. *Cognitive Linguistics, 7*(2), 183–226. doi:10.1515/cogl.1996.7.2.183

*McEnery, T., & Hardie, A.* (2012). *Corpus Linguistics: Method, Theory and Practice*. Cambridge: Cambridge University Press.

*McEnery, T., & Wilson, A.* (2001). *Corpus Linguistics: An Introduction, 2nd Ed.* Edinburgh: Edinburgh University Press.

*Pęzik, P.* (2011). Providing corpus feedback for translators with the PELCRA search engine for NKJP. In S. Goźdź-Roszkowski (Ed.), *Explorations Across Languages and Corpora: PALC 2009.* (pp. 135–144). Frankfurt am Main: Peter Lang.

*Piasecki, M., Broda, B., & Szpakowicz, S.* (2009). *A Wordnet from the Ground Up.* Wrocław: Oficyna Wydawnicza Politechniki Wrocławskiej.

*Przepiórkowski, A., Bańko, M., Górski, R. L., & Lewandowska-Tomaszczyk, B. (Eds.).* (2012). *Narodowy Korpus Jezyka Polskiego.* Warszawa: Wydawnictwo Naukowe PWN.

PWN-Oxford (2004) *English-Polish and Polish-English Dictionary.* [CD-ROM]. Warszawa: Wydawnictwo Naukowe PWN.

*Radvansky, G. A., & Zacks, J. M.* (2011). Event perception. *Wiley Interdisciplinary Reviews: Cognitive Science, 2*(6), 608–620. doi:10.1002/wcs.133.

*Rojo, A., & Valenzuela, J.* (2009). Fictive Motion in Spanish: Travellable, non-travellable and path-related manner information. In J. Valenzuela, A. Rojo, & C. Soriano (Eds.), *Trends in Cognitive Linguistics: Theoretical and Applied Models* (pp. 244–260). Frankfurt am Main: Peter Lang.

SłowoSieć 2.1 (2014). *A lexical database for Polish with mappings between Polish WordNet and English [Princeton] WordNet.* Wrocław: Wrocław University of Technology. Available at: plwordnet.pwr.wroc.pl

*Tabakowska, E.* (2003). Space and Time in Polish: The Preposition za and the Verbal Prefix za-. In H. Cuyckens, T. Berg, R. Dirven, & K.-U. Panther (Eds.), *Motivation in Language: Studies in Honor of Günter Radden* (pp. 153–177). Amsterdam: John Benjamins.

*Talmy, L.* (1996). Fictive Motion in Language and "Ception." In P. Bloom, M. A. Peterson, L. Nadel, & M. F. Garrett (Eds.), *Language and Space* (pp. 211–276). Cambridge, MA: MIT Press.

*Talmy, L.* (2000). *Toward a Cognitive Semantics, Vol. I: Concept Structuring Systems.* Cambridge, MA: MIT Press.

*Waliński, J. T.* (2013). *Complementarity of Space and Time in Distance Representations: A Corpus-based Study.* Łódź: Wydawnictwo Uniwersytetu Łódzkiego.

*Waliński, J. T.* (2014a). Atemporality of Coextension Paths. In B. Lewandowska-Tomaszczyk & K. Kosecki (Eds.), *Time and Temporality in Language and Human Experience* (pp. 103–119). Frankfurt am Main: Peter Lang.

*Waliński, J. T.* (2014b). *Time in structuring fictive motion in the National Corpus of Polish – an extended study* (COST TIMELY Research Report No. 01/2014). Lodz: University of Lodz. Retrieved on 15 Sep 2014 from http://anglistyka.uni.lodz.pl/userfiles/timely/NCP-FictiveMotionTemporality.pdf.

*Waliński, J. T.* (2014c, September 16). *Speaking Fictive Motion: Instrumentality of Coextensions Paths*. Presentation at the 'Cognitive Linguistics in the Year 2014' Conference, Częstochowa, Akademia im. Jana Długosza.

*Woelert, P.* (2011). Human cognition, space, and the sedimentation of meaning. *Phenomenology and the Cognitive Sciences, 10*(1), 113–137. doi:10.1007/s11097-010-9153-3

*Zwaan, R. A.* (2009). Time in Language, Situation Models, and Mental Simulations. In P. Indefrey & M. Gullberg (Eds.), *Time to Speak: Cognitive and Neural Prerequisites for Time in Language* (pp. 13–26). Oxford: Wiley-Blackwell.

Barbara Konat
*Adam Mickiewicz University in Poznań, Poland*

Konrad Juszczyk
*Adam Mickiewicz University in Poznań, Poland*

# Multimodal communication in career coaching sessions: lexical and gestural corpus study

**Abstract:** In the paper we aim to describe multimodal communication in career coaching sessions using recorded natural, face-to-face interactions and corpus techniques. We look on how the specific context of communication influences the lexical and gestural behaviour of speakers in the dialogues between a professional career coach and a participant. Main topic of the conversation was participant's future career. Spoken corpus of career coaching sessions was compared with standard spoken Polish corpus, to reveal four main categories of keywords: cognition verbs, words describing uncertainty and words indicating an abstract referent. Also some types and functions of gestures in career coaching were analysed with the use of NEUROGES coding system.

Keywords: multimodal communication, spoken corpus, gestures, clean coaching, coaching, neuroges

## 1. Introduction

Recent decades have witnessed major technological and theoretical changes in the study of human communication. Firstly, new technologies allow modern researchers to record communicative interactions in all modalities and to store large amounts of data (hundreds of hours of audio-visual recordings, numerous participants, millions of words and gestures). Secondly, the theoretical shift in communication sciences (including linguistics, psychology, sociology, anthropology) has changed the researchers' focus: from verbal communication to the whole spectrum of multimodal interactions. These two changes constitute the current paradigm in natural human communication research.

Staying within the scope of this paradigm, in this paper we analyse how the specific context of communication influences the lexical and gestural behaviour of speakers.[1] We aim to describe multimodal communication in career coaching sessions using recorded interactions and corpus techniques. We also adopt a broad view of communication: following Hymes (1974), we claim that who speaks to whom, about what and in what circumstances, influences lexical and gestural choices. Will speakers in career coaching sessions use specific words and gestures?

Multimodal communication is a broad term. Allwood, in his studies, paid attention to the ways in which various modalities influence each other. He also pointed out that studies of multimodal communication and pragmatics should be interconnected (Allwood, 1995). In a more recent study, Bonnachi & Karpiński (2014) claimed that "verbal language, gestures, facial expressions, voice and movements are regarded as mutually interdependent" and they "interplay in the rising of meaning". The definition of multimodal communication we propose here follows this tradition in that it encompasses a broad view of multimodality. In our view, human face-to-face communication is a dynamic, interactive and multimodal process. Such communication is multimodal, as people use various means to communicate. So it is due to the limited scope of this paper, not our theoretical approach, that only two modalities will be analysed: lexical and gestural.

Analysis of multimodal communication will be presented here using the example of career coaching sessions. The International Coach Federation defines coaching as "partnering with clients in a thought-provoking and creative process that inspires them to maximize their personal and professional potential".[2] Usually, the complete coaching process consists of 6 to 10 sessions, each lasting between 30 and 90 minutes. A coaching session can be considered as multimodal interaction, where the coach, using specific questions, repetitions and paraphrases, helps the client (also named *coachee*) to reflect on his or her professional or personal experience, plans and career. Coaching sessions share many qualities with everyday dialogues, as they are all direct, multimodal face-to-face interactions. What is distinct in coaching sessions is that they are support-oriented and engaging for both coachee and coach. Coaches are responsible for the session's structure, while clients focus on their reflections and results. Moreover, coaches are trained to maintain a rapport and give confidence to clients through nonverbal behaviour (for example, by mirroring and mimics).

1 The paper has been prepared within a research project funded by the Polish National Science Centre (decision DEC- 2011/03/D/HS6/05993). Authors would like to thank all team members and subjects of the study.

2 [@]: http://www.coachfederation.org/need/landing.cfm?ItemNumber=978&navItemNumber=567 [date of access: 15.09.14]

## 2. Methods and materials: The Multimodal Career Coaching Sessions Corpus

The Multimodal Career Coaching Sessions Corpus (henceforth: MM or MULTIMET) used in this paper is a corpus of recorded dialogues between a professional career coach and a participant. The participants were 45 volunteers, aged 25–35 years (15 males and 30 females) – graduate and doctoral students and alumni from various schools in Poznań. Each participant took part in a minimum of two career-coaching sessions (one participant took part in 6 sessions). In total, 96 clean coaching sessions were recorded between April and July 2013, with cameras and microphones in the studio organized at Adam Mickiewicz University in Poznań. The average session length was 40 minutes. Data was stored and prepared for further stages of analysis: gesture annotation and transcription, and a multimodal corpus was created.

Coaching sessions were conducted according to the Polish version of Clean Language (Pieśkiewicz & Kołodkiewicz, 2011). Clean Language is a method used worldwide in coaching and psychotherapy, the main assumption of which is that people will use metaphors and symbols as a way to describe difficult experiences and concepts, such as: the future, development and job-seeking (Sullivan & Rees, 2008). This technique consists of twelve questions designed to develop metaphors and symbols in dialogue. Questions are called *clean*, because the coach uses exact wording (not a paraphrase, but rather a *parrotphrase*) of a coachee's answer to the previous question. In the Clean Coaching scenario used in our study, the coach starts with a question about career and after that asks for a comparison. When the comparison of a concept is established, the coach asks about the details of the career metaphor, such as its size, shape and colour. In order to help the coachees organise their thoughts, the coach usually repeats parts of the coachee's utterances and relates them to specific questions. The short sample of a career Clean Coaching session presented in Table 1 shows the clean question, the answer and another clean question with repetitions in bold.

*Table 1: Example of an opening question and questions with repetitions in a Clean Coaching session*

| Speaker | Utterance (and its English equivalent) | Comments |
|---|---|---|
| COACH | *-kiedy myślisz o swojej karierze to ta kariera jest jak co*<br>[when you think about your career this career is like what] | opening question |
| COACHEE | ***-fale***<br>**[waves]** | answer |

| Speaker | Utterance (and its English equivalent) | Comments |
|---|---|---|
| COACH | *-jakie są te* ***fale***<br>[what are these **waves**] | question about attributes |
| COACHEE | *-wzburzone energiczne ale piękne takie pełne życia coś w tym stylu* ***wolne przestrzeń*** *właśnie* ***wolna*** *może jeszcze*<br>[vigorous rough but beautiful full of life something like that **free space** yes free maybe in this way] | more detailed answer of coachee |
| COACH | *-te* ***fale przestrzeń*** *co jeszcze*<br>[these **free waves spaces** what else can you say about that] | further question about attributes |

The dialogues were transcribed in ELAN software. As a basic transcription unit, an intonation phrase was used (see Szczyszek, 2013 for details of this transcription method). No capital letters or punctuation were added to the transcription. Almost 24 hours of dialogues were transcribed, which corresponds to 40% of the total. For the current analysis, only participants' utterances from the corpus were used, thus leaving a corpus of a size of 99 251 words. The utterances of the coach – mainly questions and comments – were removed, as they were predetermined by the Clean Coaching scenario. The transcribed corpus consists of the speech of 23 participants, 18 female and 5 male. Each participant contributed between 1 000 and 12 000 tokens to the corpus.

The annotation of gestures was done using the NEUROGES coding system (Lausberg, 2013) with the aid of ELAN software (Lausberg & Sloetjes, 2009). The NEUROGES coding system was developed by Hedda Lausberg's team, originally to identify features of hand movements related to neurological diseases. The assumption behind the NEUROGES coding system is that hand movements are related not only to language processes or linguistic units and rules, but also to general cognitive, emotional and interactive processes. The annotation scheme used in our project is based on the NEUROGES manual, with some minor changes. Four NEUROGES categories of hand movements were used: Structure, Focus, Function and Type. Only those hand movements that have a stroke phase are considered gestures. These include: hand movements that are phasic and repetitive (in terms of their Structure), and hand movements performed in space or on objects (in terms of their Focus). In the first stage of gesture annotations, trained annotators worked on recordings of sessions with muted sound, so they relied only on kinesic features of hand movements. Sessions were annotated on four layers: both hands separately for the coach and coachee. In the next stage, interpretation of hand movements was conducted together with categorisation in terms of Functions and Types.

# 3. Results

## 3.1 Lexical characteristics of coaching sessions

To reveal the specific words used during coaching sessions, the Multimodal Career Coaching Session Corpus (MM) was compared with a sample from the PELCRA spoken corpus and a keyword list was generated.

PELCRA spoken is a publicly available corpus of Polish spoken data (Pęzik, 2012). From the PELCRA corpus, a 2 million sub-corpus of conversational data was extracted (leaving aside transcripts of mass media communication and parliamentary sessions). From this 2 million corpus, a 100 000 token sample was created to match the size of the MM corpus. The sample was taken with a systematic method: the corpus was divided into 200 parts and every 20th part was taken, creating the PELCRA Sample corpus (PS).

The MM corpus was compared with the PELCRA Sample corpus (PS). A keyword list was generated with the use of AntConc software (Laurence, 2009). Three main categories of keywords were identified: cognition verbs, words describing uncertainty and words indicating an abstract referent. These categories show the difference between career coaching sessions (MM corpus) and standard spoken Polish (PELCRA sample).

The first category consists of cognition verbs, that is verbs describing mental states, desires or feelings. Frequency of those verbs is different in various genres of Polish language (Kokorniak&Fabiszak, 2014). The career coaching corpus indicated frequent use of four of them: *wiem* [I know], *myślę* [I think], *chcę* [I want], *czuję* [I feel]. All verbs were uttered in first person singular present tense, hence referring to the speaker's own mental states, as experienced at the moment of speaking[3]. Figure 1 below presents differences in the frequency of those four verbs in the career coaching sessions corpus (MM) compared to the sample of standard spoken Polish (PS). Next, Table 2 presents Log Likelihood values, which show statistically significant differences between the frequency of use of cognition verbs (99.99th percentile; 0.01% level; $p < 0.0001$; critical value = 15.13). To ensure higher validity, results were re-calculated with the complete 2M PELCRA conversational corpus, confirming statistically significant differences. This means that words describing mental states are overused in career coaching sessions in comparison to standard spoken Polish.

3 Polish is an inflectional language, where an overt subject is not required since person and number are indicated in the suffix: wiedzieć (INF)-wiem (1SG), myśleć (INF)-myślę (1SG), chcieć (INF)-chcę (1SG), czuć (INF)-czuję(1SG). Specifics of Polish corpus linguistics were described by Lewandowska-Tomaszczyk (2005).

*Figure 1. Frequency of word tokens from the category of cognition verbs in the career coaching sessions corpus (MM) and the 100 000 token PELCRA Sample corpus (PS).*

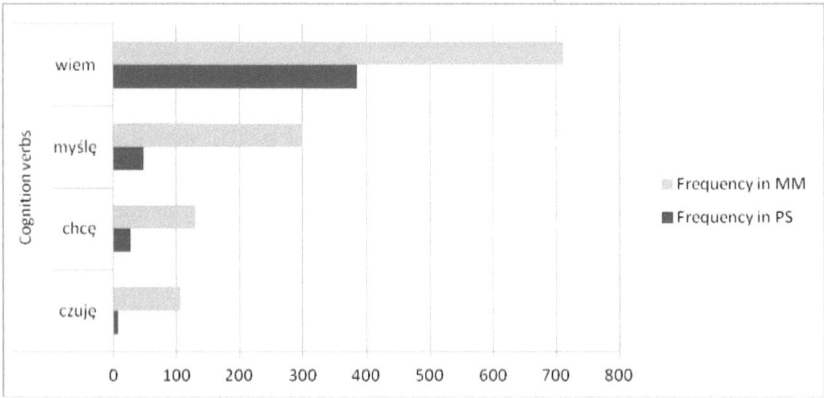


*Table 2: Keywords from the category of cognition verbs and their English equivalents with the frequency in the career coaching sessions corpus (MM) and the 100 000 token PELCRA Sample corpus (PS). The Log Likelihood value shows statistically significant overuse of those words in the career coaching sessions corpus*

| Word (Polish) | English equivalent | Frequency in MM | Frequency in PS | LL value |
|---|---|---|---|---|
| *wiem* | I know | 712 | 386 | 91.979 |
| *myślę* | I think | 300 | 48 | 198.295 |
| *chcę* | I want | 131 | 27 | 72.507 |
| *czuję* | I feel | 106 | 7 | 102.223 |

Cognition verbs are used in coaching sessions in various contexts. They can provide information about lack of knowledge (example 1) or needs and feelings (examples 2, 3 and 4):

1) *no **nie wiem** brakuje takiego takiej chęci może **nie wiem** ryzyka*
   [**I don't know** what is missing—some some willingness of **I don't know** risk] (66-S1)[4].
2) ***myślę** że chciałabym żeby rzeczy szły do przodu szybciej*
   [**I think** that I would like things to go faster] (90-S2).

4 The number indicates speaker number and coaching session number: 91-S1 means speaker 91, during their first coaching session.

3) *dobrze mi jest tak jak teraz idę i* ***chcę*** *utrzymać ten rytm* ***nie chcę nie chcę*** *już wracać*
   [I'm fine, walking just as I do now and **I want to** keep this rhythm **I don't want to not want to** go back](91-S2).
4) *ja mówię że to jest bardzo długa droga* ***czuję*** *to że to jest długa droga*
   [I'm saying that it is a very long road **I feel** it is a long road] (91-S1).

The second category of keywords includes those indicating the speaker's uncertainty. Words from this category may function as parentheses or hedges, indicating that the speaker is not ready to take full responsibility for what is going to be said. Keywords included in this category are: *właśnie* [exactly], *może* [maybe], *gdzieś* [somewhere], *jakby* [somewhat, somehow], *wydaje (mi się)* [it seems (to me)], *(w) sensie* [(in the) sense], *powiedzmy* [let's say]. Figure 2 below presents differences in the frequency of those words in the career coaching sessions corpus (MM) compared to the sample of standard spoken Polish (PS) and Table 3 presents Log Likelihood values, which show statistically significant differences.

*Figure 2. Frequency of word tokens from the category of uncertainty words in the career coaching sessions corpus (MM) and the 100 000 token PELCRA Sample corpus (PS).*

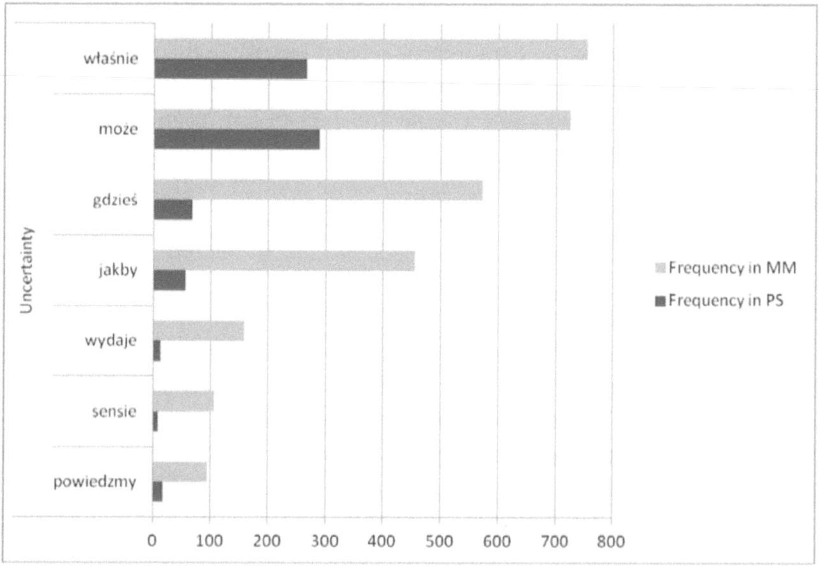

*Table 3: Keywords from the category of uncertainty and their English equivalents with their frequency in the career coaching sessions corpus (MM) and the 100 000 token PELCRA Sample corpus (PS). The Log Likelihood value shows statistically significant overuse of those words in the career coaching sessions corpus*

| Word (Polish) | English equivalent | Frequency in MM | Frequency in PS | LL value |
|---|---|---|---|---|
| *właśnie* | exactly | 755 | 267 | 233.325 |
| *może* | maybe | 725 | 289 | 185.272 |
| *gdzieś* | somewhere | 572 | 67 | 447.088 |
| *jakby* | somehow | 456 | 56 | 348.498 |
| *wydaje* | it seems | 159 | 12 | 147.291 |
| *sensie* | sense | 106 | 8 | 98.194 |
| *powiedzmy* | let us say | 94 | 16 | 59.730 |

Uncertainty words in the corpus of career coaching sessions (MM) are often used in the context of describing future jobs or possible career paths as presented in examples 5–11 below. Sometimes they are used repeatedly, one after another, especially when claims included in the utterance may be considered controversial or unpopular, just as in example 8.

5) *ale no to nie musi być coś* ***właśnie*** *związanego z zawodami medycznymi*
   [But it does not have to be something related **exactly** to the medical profession] (02-S1).
6) ***może*** *być mocny impuls który wywoła mniejszą reakcję ale* ***może*** *być malutki impuls który wywoła ogromną reakcję*
   [**Maybe** a large impulse will come and cause a small reaction but **maybe a** smaller impulse will cause a larger reaction] (66-S1).
7) *to mogą być właśnie takie* ***gdzieś tam*** *zadania które* ***gdzieś tam*** *wykonywaliśmy*
   [There could be just tasks **somewhere** tasks we were completing **somewhere]** (75-S2)
8) *bo dla mnie jakieś* ***jakby*** *pewne konwenanse związane z pracą* ***jakby*** *miejsce czas są pewnym ograniczeniem i z tym* ***jakby*** *nie czuję się dobrze*
   [Because for me some **somehow** certain conventions connected to work **somehow** time and space are a certain restriction and I don't **somehow** feel good about that] (02-S1).
9) *także myślę że to jest no nie wiem* ***wydaje mi się że*** *jestem teraz w takim momencie*
   [So I think it is oh I don't know **it seems to me** that I'm in such a moment right now] (75-S1).

10) *znam takie osoby które w ogóle nie usłyszały dobrego słowa* ***w sensie*** *znaczy od rodziców*
[I know some people who never really heard a good word **in the sense** I mean from their parents] (54-S1)

11) *pracę która będzie dla mnie satysfakcjonująca ale też bezpieczna* ***powiedzmy*** *w pewien sposób*
[Work that will be satisfactory for me, but at the same time **let's say** safe in some way] (02-S1).

The third category encompasses words with an abstract referent. By abstract we understand here non-material or non-literal. This category includes: *czas* [time], *rzeczy* [things], *życie* [life], *możliwości* [possibilities], *praca* [work], *celu* [aim]. Figure 3 below presents differences in the frequency of those words in the career coaching sessions corpus (MM) compared to the sample of standard spoken Polish (PS) and Table 4 presents Log Likelihood values, which show statistically significant differences.

*Figure 3. Frequency of word tokens from the category of words with an abstract referent in the career coaching sessions corpus (MM) and the 100 000 token PELCRA Sample corpus (PS).*

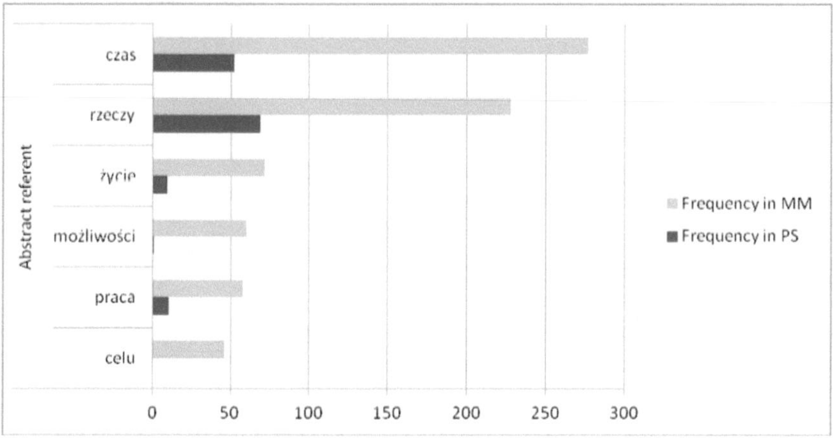

*Table 4: Keywords from the category of words with an abstract referent and their English equivalents with their frequency in the career coaching sessions corpus (MM) and the 100 000 token PELCRA Sample corpus (PS). The Log Likelihood value shows statistically significant overuse of those words in the career coaching sessions corpus.*

| Word (Polish) | English equivalent | Frequency in MM | Frequency in PS | LL value |
|---|---|---|---|---|
| *czas* | time | 277 | 52 | 164.540 |
| *rzeczy* | things | 228 | 69 | 86.652 |
| *życie* | life | 72 | 9 | 54.552 |
| *możliwości* | possibilities | 60 | 1 | 73.208 |
| *praca* | work | 58 | 10 | 36.543 |
| *celu* | aim | 46 | 0 | 62.872 |

The presence of abstract referents was pre-determined by the topic of conversation and the method of Clean Coaching: career and life plans, possibilities and goals, things to do:

12) *że praca im daje takie poczucie właśnie że są potrzebni że ten **czas** który tutaj mają nie jest zmarnowany*
[That this work gives them the feeling that they are useful, that their **time** here is not wasted] (80-S2).
13) *tylko te **rzeczy** które mnie rozwijają które do czegoś mi są potrzebne staram się realizować*
[I'm trying to do only those **things** that can help me to develop those that I need] (54-S1).
14) *to się okaże to wszystko zależy od tego jak potoczy się moje **życie***
[We will see it all depends on how my **life** goes] (29-S2).
15) *pojawiają się nowe pomysły i nowe **możliwości***
[New ideas and new **possibilities** appear] (60-S3).
16) *w sensie żeby **praca** nie była całym życiem bo to nie bo to nie jest fajne*
[In the sense that your **work** is not your whole life because this is not this is not cool](66-S2).
17) *szukam cały czas jakiegoś swojego konkretnego **celu** to co chciałabym robić tego co lubię*
[I am searching all the time for my own specific **aim** something that I would like to do something I like] (21-S2).

To sum up: words describing mental states, words indicating uncertainty and words with an abstract referent are more frequent in career coaching sessions than in typical spoken Polish, and this difference is statistically significant. Therefore, the high frequency of the words analyzed above may be considered a distinct feature of career coaching sessions, setting them apart from standard spoken Polish.

The three categories identified among keywords in Study 2 are not clear-cut, as the words listed in them have more than one function. In the example 1) we can observe how the cognition verb "know" also serves as a softener, a marker of uncertainty. However, the presented categories seem to be sufficient to identify differences between standard conversations and career coaching sessions.

## 3.2 Types and functions of gestures in career coaching

Keywords in coaching sessions are accompanied by co-speech gestures. Below we present analysis just of four of them, related to the concepts of time, vision and career management. Examples are presented on screenshots from ELAN, transcription, translation and frames. Figure 4 below presents a screenshot from the ELAN software with the gesture description. The first and second tiers include a transcription of the coachee's words. The first tier shows complete utterance, whereas the second shows the utterance segmented into words so temporal alignment with hand movement is visible. Labels used in description are taken from NEUROGES coding system. First label – *phasic in space* or *repetitive in space* refer to structure and focus step in annotation. Second label – *emotion-fist clenching* and others refer to function and type step in annotation.

The first example of gesture identified in a coaching session is a co-speech gesture: phasic in space—emotion-fist clenching.

18) *po tym* ***czasie*** *może być może już po tym egzaminie teraz ale po tym czasie obrony na pewno chcę właśnie żwawo ruszyć żeby ten mój największy cel osiągnąć*
[After that **time** maybe after that exam now but after that time of the defence exam for sure I want to start quickly to achieve my biggest aim].

Figure 5 presents photos of this gesture. In Figure 1, Lines 3–5 describe the gesture. The coachee performs a kind of emblematic gesture with her right hand. This gesture starts just before the verb *chcę* (*I want*) and its stroke phase lasts until the word *żwawo* (*quick*). In NEUROGES terms it is *phasic in space*, because the movement of the hand is done in one trajectory. The movement in the stroke phase reflects the grasping of an imaginary object and is finished with *fist clenching*. This type of hand movement can be classified in NEUROGES function terms as an *emotion/attitude* function or a *motion quality presentation*.

*Figure 4. ELAN software screenshot presenting a co-speech gesture description in the NEUROGES system for the gesture accompanying example 18.*

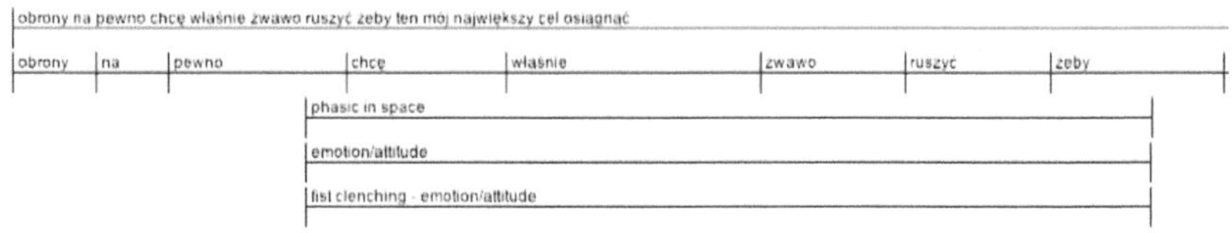


*Figure 5. The gesture accompanying example 18 – fist clenching. The arrows are used to show the direction of hand movement.*

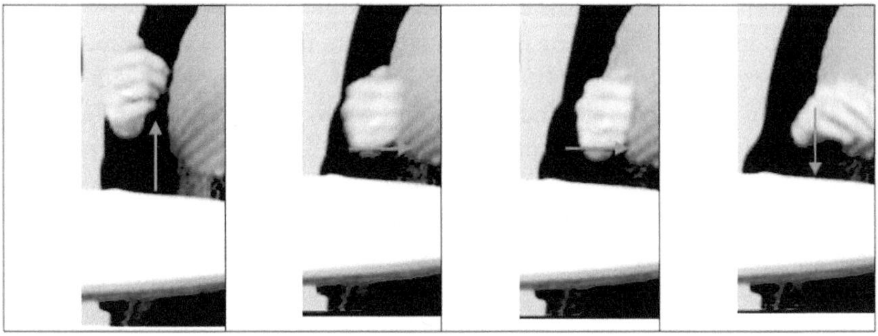

The second gesture is a *repetitive in space–spatial relation representation*. This co-speech gesture is performed as an accompaniment to the utterance:

19) *muszę miarkować ten* ***czas*** *pomiędzy te zawodowe kwestie i jeszcze też wiadomo prywatne i te szkolne i w ogóle ten mój cel*
[I need to manage this **time** between professional issues and you know private and school-related and in general this my aim]

Figure 6 presents photos of the gesture. Through this utterance, the coachee performs an emblematic gesture with her right hand. Since the movement is repeated several times in a similar, vertical trajectory, it would be called *repetitive in space,* in NEUROGES terms. The right hand is poking the left hand, which is flat. The combination of fingers in the right hand is "1–5 touching", which means that all fingers are joined together, pointing downwards (Bressem 2013). Moreover, both hands move from the body-midline to the left side of the speaker. Every poke is indexing each item listed in the coachee's utterance: professional issues, private and school-related issues and the aim. This gesture is a way of structuring the representation of the coachee's domains of time management. Each domain has its place in the centre of the gesture space (see McNeill, 1992 for gesture space definition). The coachee is pointing to each of the places and shows the spatial relations between them. This can be treated as a means of delimitation of those domains. In NEUROGES terms, on a function/type level, this gesture may be categorised as *spatial relation presentation-position.*

*Figure 6. The gesture accompanying example 19 – repetitive in space gesture-spatial relation representation. Frames are shown from left to right.*

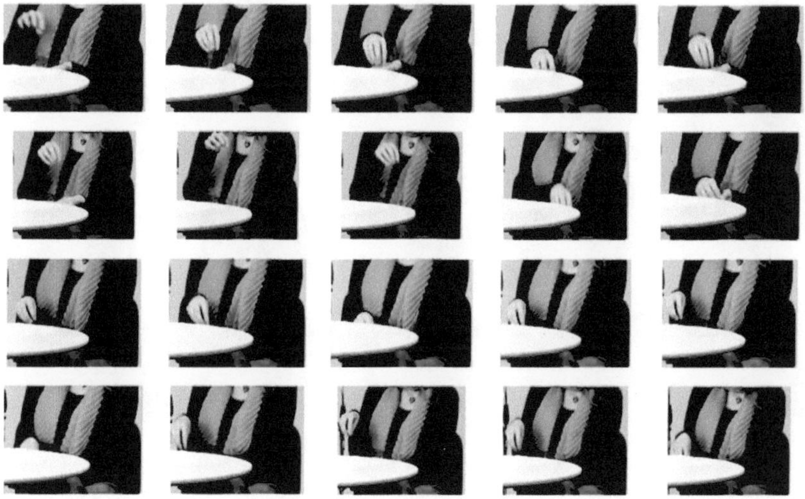

The third example of a gesture is a *repetitive in space–motion quality presentation.* Figure 7 below presents a screenshot from the ELAN software with the gesture description. The first and second lines include a transcription of the coachee's words:

20) *który mi się tam gdzieś ta* ***wizja*** *kształtuje ale na razie się*
'a **vision** of this is shaping for me somewhere but for now'

*Figure 7. The gesture accompanying example 20 – repetitive in space-motion quality presentation. Frames are shown from left to right.*

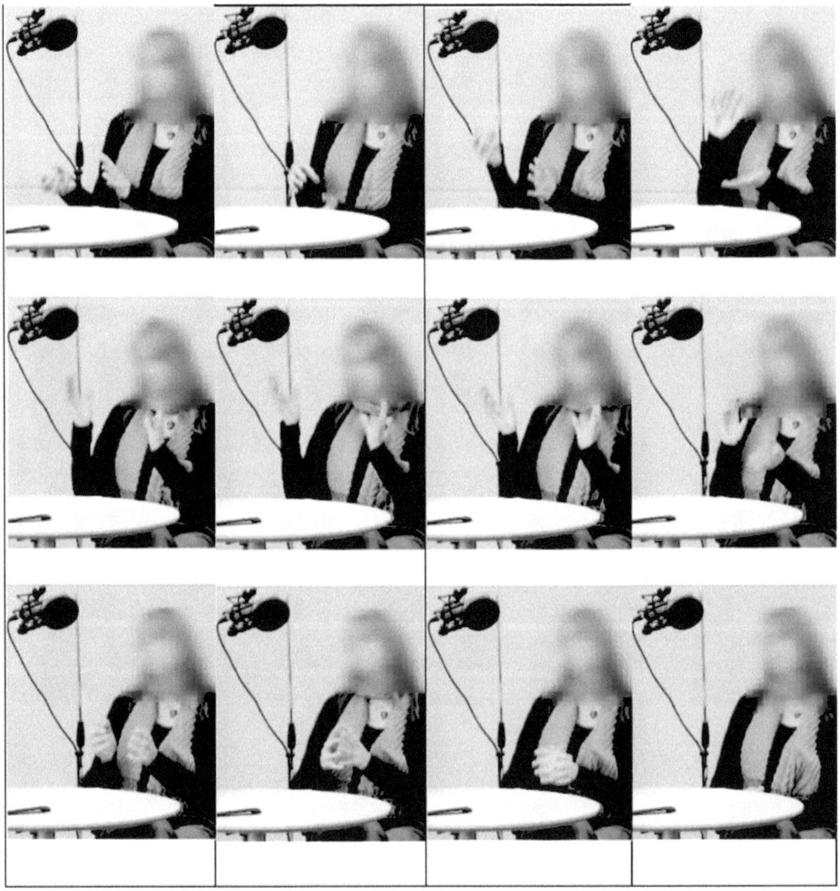

Figure 8 presents photos of the gesture. The hand movement performed while speaking about the shaping of the vision is bimanual (both hands are active), simultaneous and alternating (as one hand goes up, the other goes down and then vice versa) rotation of an imaginary object. In NEUROGES terms it is *repetitive in space* on a structure/focus level and *motion quality presentation-manner* on a function/type level.

*Figure 8. ELAN software screenshot presenting the co-speech gesture description in the NEUROGES system for the gesture accompanying example 20.*

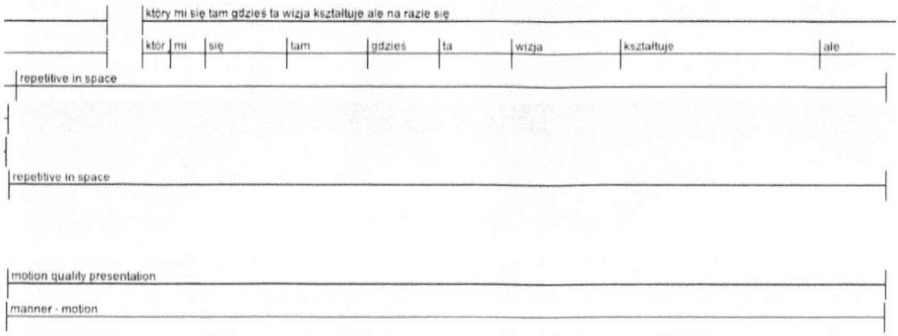


The fourth example gesture is *phasic in space–egocentric deictic-external target*. Figure 9 below presents a screenshot from the ELAN software with the gesture description. The first and second lines include a transcription of the coachee's words:

21) *zostawiam ją po prostu aż uporządkuję sobie*
[I'll leave that until I order (my things)]

Figure 10 presents photos of the gesture. The gesture is used while speaking about the future, and it is *phasic in space* on a structure/focus level and an *egocentric deictic-external target* on the function/type level. It is performed with two hands joined and moved forward, from the speaker's body-midline.

*Figure 9. ELAN software screenshot presenting a co-speech gesture description in the NEUROGES system for the gesture accompanying example 21.*

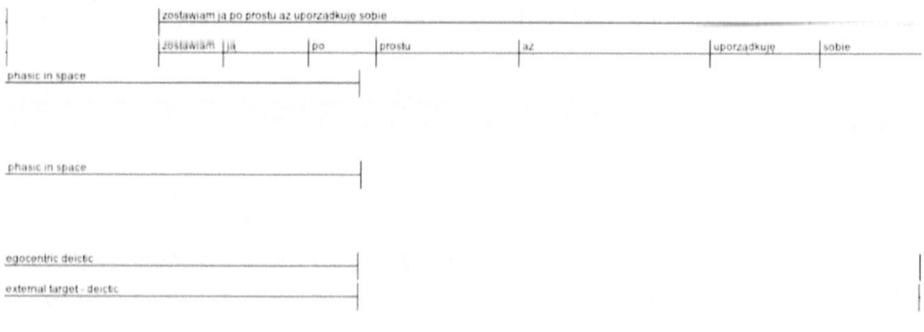

*Figure 10. The gesture accompanying example 21 – phasic in space – egocentric deictic-external target. Frames are shown from left to right.*

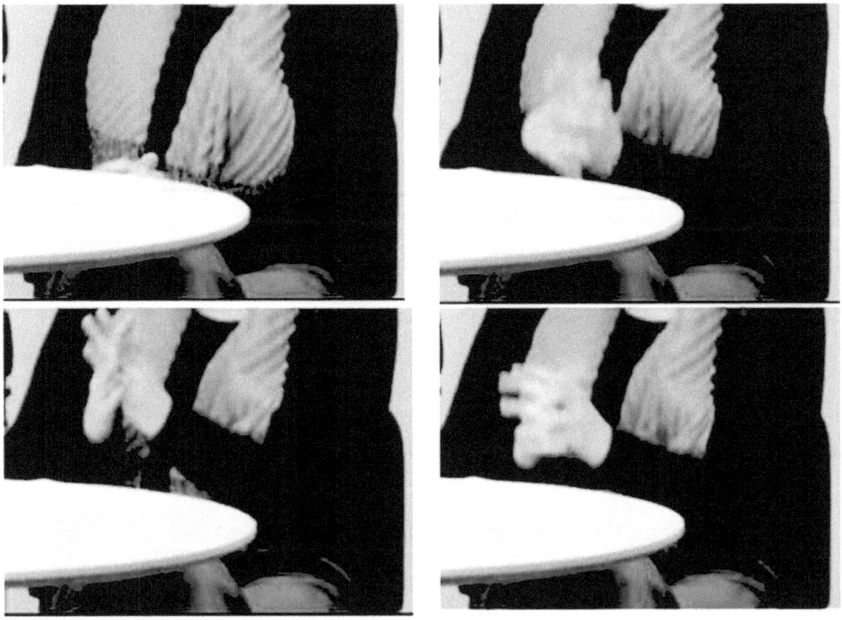

This short study on selected gestures from career coaching corpora serves as an example of spontaneous hand movements used to enrich interaction and embody the concepts behind the coachee's words. Although it is not easy to decide which word or concept is linked directly to a particular hand movement, we can see that these hand movements are not accidental. Moreover, in some cases, some gestures are repetitively used alongside utterances.

## 4. Discussion

Analysis of the multimodal communication corpus in career coaching sessions showed that speakers use specific gestures and words to talk about abstract concepts such as time, visions of the future, careers and life goals. Speakers, when talking about abstract referents, show a high level of uncertainty and use cognition verbs to refer to their own states. Their gestures and lexical choices are motivated and influenced by the context of the conversation. Speakers use both words and co-speech gestures to shape complex, career-related ideas.

Further investigation of the collected material is required, which will include other modalities and combine gestures and words in quantitative studies. Examples of gestures chosen for this paper are also used by various coachees in other sessions collected in MM corpus. We expect to find more instances of functions and types of hand movements and show temporal alignment of them with specific words used with them. Moreover, more detailed analyses of patterns in hand movements and word classes are planned. Such qualitative and quantitative analysis may show what kind of hand movements (in NEUROGES terms on Structure/Focus and Function/Type level) goes with specific words in many sessions with different coachees and coaches.

The specific context of career coaching sessions, abstract referents and the support-oriented structure, influence speaker's gestures and lexical choices. Words and gestures in natural, face-to-face communication are interdependent in the creation of meaning. However, research is needed to obtain more representative results. We hope to fulfil this need soon.

## References

*Allwood, J. S.* (1995). *An activity based approach to pragmatics* (Vol. 76). University of Gothenburg.

*Bonacchi, S., & Karpiński, M.* (2014). Remarks about the use of the term "multimodality". *Journal of Multimodal Communication Studies*, (1), 1–7.

*Bressem, J.* (2013). A linguistic perspective on the notation of form features in gestures. [in:] Mueller, Cienki, Fricke, Ladewig, McNeill, Teßendorf (eds.) 2013, *Body – Language – Communication* (HSK 38.1), de Gruyter, 1079–1098.

*Hymes, D. H.* (1974). *Foundations in sociolinguistics*. University of Pennsylvania Press.

*Kokorniak, I. M. Fabiszak* (2014). Grammaticalization of Polish mental predicate prefixes in: Sylvie Hancil, Ekkehard König (Ed.), *Grammaticalization–Theory and Data*. Amsterdam: John Benjamins.

*Laurence, A.* (2009). Issues in the design and development of software tools for corpus studies: The case for collaboration. In P. Baker (Ed.), *Contemporary Corpus Linguistics*. London; New York: Bloomsbury Academic.

*Lausberg, H.* (2013). *Understanding Body Movement. A Guide to Empirical Research on Nonverbal Behaviour With an Introduction to the NEUROGES Coding System*. Frankfurt am Main: Peter Lang.

*Lausberg, H., & Sloetjes, H.* (2009). Coding gestural behavior with the NEUROGES-ELAN system. *Behavior Research Methods, 41*(3), 841–849.

*Lewandowska-Tomaszczyk, B.* (2005). *Podstawy językoznawstwa korpusowego* [The basics of corpus linguistics]. Łódź: Wydawnictwo Uniwersytetu Łódzkiego.

*McNeill, D.* (1992). *Hand and Mind: What Gestures Reveal about Thought*. University of Chicago Press.

*Pęzik, P.* (2012). Wyszukiwarka PELCRA dla danych NKJP [PELCRA search engine for the National Corpus of Polish]. In A. Przepiórkowski, M. Bańko, R. Górski, & B. Lewandowska-Tomaszczyk (Eds.), *Narodowy Korpus Języka Polskiego*. Wydawnictwo Naukowe PWN.

*Pieśkiewicz, B., & Kołodkiewicz, M.* (2011). Metafora w coachingu: zastosowanie metody Clean Coaching w pracy z organizacjami [A metaphor in coaching: using Clean Coaching method in organisational coaching]. In L. D. Czarkowska (Ed.), *Coaching. Katalizator rozwoju organizacji*. Warszawa: New Dawn.

*Sullivan, W., & Rees, J.* (2008). *Clean Language: Revealing Metaphors and Opening Minds* (1st edition edition.). Bethel, Conn: Crown House Publishing.

*Szczyszek, M.* (2013). *O porozumiewaniu się międzyludzkim. Wariant mówiony języka polskiej wspólnoty komunikatywnej: słowotwórstwo, składnia, leksyka.* [On human communication. Spoken variety of Polish communicative community: word formation, syntax and lexis]. Poznań: Wydawnictwo Rys.

Victoria Kamasa
*Adam Mickiewicz University in Poznań, Poland*

# Research design in corpus-supported critical discourse analysis

**Abstract:** The usage of corpus linguistics for critical discourse analysis has become increasingly popular in the last decades. In this paper we use the perspective of research process design to present some general tendencies in corpus-supported critical discourse analysis on the basis of over 30 journal articles. We reconstruct the way CDA researchers who use corpus techniques deal with research design starting with research questions and nominal definition through operationalization and sampling and finishing with the presentation and interpretation of the results. The overview shows that the overall approach to use corpus methods in CDA remains very similar. It also reinforces the usefulness of study design description as both a tool for planning one's own research and an analytic instrument appropriate for observing the main empirical tendencies in a research field.

Keywords: critical discourse analysis, corpus-assisted discourse studies, corpus linguistics, research design, concordances, collocates, keywords.

## 1. Introduction

The usage of corpus linguistics (CL) for critical discourse analysis (CDA) has become increasingly popular in the last decades. According to the Scopus database, the total number of publications using this approach in 1990's came to 3, in 2000's it was 29, and since 2010 it has already reached 47[1]. This tendency is also visible in the leading conferences in the field: during the 2014 Critical Approaches to Discourse Analysis Among Disciplines (CADAAD) conference almost 40 authors used some form of corpus analysis in their studies. Theoretical reflection on the use of CL in CDA is also growing: Baker (2006) offers an exhaustive overview of most corpus techniques applicable in different forms of discourse analysis. Some authors such as Degano (2007), Lukac (2011), and Prentice (2010) focus on the advantages of corpus approaches for CDA, whereas Baker et al. (2008) compare and contrast results obtained by qualitative CDA with corpus-supported analysis.

1 Presented numbers result from search term: "KEY ("critical discourse analysis" ) AND TITLE-ABS-KEY (corpus OR corpora OR "corpus linguistics")" (Scopus, 2014).

Therefore, the application of CL for CDA might be considered one of the current directions in this field.

In this paper we use the perspective of research process design to present some general tendencies in corpus-supported[2] critical discourse analysis (henceforth: CL/CDA). The current overview is based on over 30 journal articles published between 2002 and 2014. These papers have been chosen on the basis of two criteria:

- Declarative commitment of the authors to the CDA framework (or one of the frameworks within CDA such as discursive-historical analysis);
- Usage of at least one corpus linguistic techniques to answer the research question.

As an analytical tool we will use the stages of research process described by Babbie (2010). Babbie's concept will be slightly modified to fit specifically into research related to language rather than general social research. We will therefore reconstruct the way CDA researchers who use corpus techniques deal with each stage of the research process.

## 2. Stages of research design

In numerous publication concerning planning and conducting research in social sciences, the research process is divided in various stages and substages. For example, Babbie (2010) describes eight stages, the first being the formulation of the research problem and the last being the analysis. In comparison, Babiński (1997) distinguishes as many as 12 stages. On the other hand, Frankfort-Nachmias et al. (2001) concentrate on the most crucial elements of this process, which in their opinion are research problem, variables, relations and hypothesis. Drawing on the above mentioned models, we suggest here to see the stages of research design as a series of 10 questions that must be answered in order to perform research and present results:

1. What do we want to know (research problem)?
2. What is our theoretical background?
3. How we define key notions (nominal definitions)?
4. What are we interested in? What does it depend on? (dependent and independent variables?

2 The distinction between corpus-supported, corpus-based and corpus-driven analysis are not widely discussed among CDA researcher. Therefore, in this paper we will use "corpus-supported" as an umbrella term for all these approaches.

5. How can we see the subject of our research (operational definitions)?
6. What methods do we use?
7. What data do we analyze?
8. How do we choose the data we analyze?
9. How do we present the results?
10. How do we interpret the results?

These questions will be used in the following paragraphs to discuss the way CL is used to support CDA.

## 3. Research design in corpus-supported CDA

### 3.1 Research questions

The most common research question formulated by CL/CDA authors concerns the discursive construction of X in the source Y. Studies therefore aim at describing discourses related to some social phenomena (X) in a collection of texts (source Y)[3]. An example of such phenomena might be social groups as in the case of Gabrielatos and Baker (2008), who studied how refugees and asylum seekers are discursively constructed in British press articles. CL/CDA research might also concentrate on events which are widely discussed in public, such as same-sex relationships (Bachmann, 2011), friendly fire accident (Albakry, 2004) or in vitro fertilization (Kamasa, 2013). Other studies might be related to ideologies, for example sexism (Mohamad, Subakir Mohd Yasin, Bahiyah, Yuen, Zarina, & Azhar, 2012), nationalism (Edwards, 2012) or appreciation of eating-disorders (Lukac, 2011). Last but not least, some global processes are addressed: Lischinsky (2011) is interested in the construction of crises, whereas Forchtner and Kolvraa (2012) investigate the discursive construction of *Europe*.

Less frequently, the analysis concerns some characteristics of chosen discourse. For instance, Baker (2006) concentrates on features of pro- and anti-foxhunting voices in parliament debate and Kamasa (2014) characterizes the discourse of Polish Catholic Church on in vitro fertilization. In such cases the scope of attention is moved from the construction of a particular phenomenon to general discursive features.

3 The problem of selection of the texts the research is based on will be described in one of the following section (sampling).

### 3.2 Theoretical background

Authors of all papers commit to some form of CDA. CDA is seen as a movement which is both recognizable from the "outside", as having common features, and self-aware, in the sense that its representatives believe themselves to be working within a "critical" paradigm as far as discourse analysis is concerned (Wodak, 2011). This movement aims at a "detailed description, explanation and critique of the ways dominant discourses (indirectly) influence (…) socially shared knowledge, attitudes and ideologies" (van Dijk, 1993, p. 258). The second basic feature of the CDA framework is taking (if possible) the perspective of those who suffer from dominance and inequality (van Dijk, 1993) and focus "on the role of discourse in the production and reproduction of power abuse or domination" (van Dijk, 2001, p. 96).

Nevertheless the level of commitment to these basic assumptions of CDA varies throughout the papers. Whereas in case of analysis related to the presentation of immigrants (Gabrielatos & Baker, 2008; O'Halloran, 2009) in British press, same-sex marriage in parliamentary debate or sexism in school textbooks (Mohamad, Subakir Mohd Yasin et al., 2012) both the concentration on power-relation and taking the side of the disprivileged is clearly visible, analysis concerning Scottish (Prentice, 2010) or Quebéc (Freake, Gentil, & Sheyholislami, 2010) national identity seem to be less involved in the discussion of the domination and discrimination.

### 3.3 Nominal definitions

The basic notion for the discussed studies is clearly discourse. It is defined in the tradition of CDA as one of "many representations of social life" (Fairclough, 2001, p. 125). The discourse is therefore described and understood as "a set of meanings, metaphors, representations, images, stories, statements and so on that in some way together produce a particular version of events surrounding any one object, event, person etc." (Burr, 1995, p. 32). Nevertheless, the majority of the analyzed papers pay little attention to the definition of discourse, understanding of this concept seems rather to be taken for granted.

### 3.4 Variables

The term variables is here understand and used in far less orthodox way than in traditional quantitative studies. In the majority of the papers the factors considered in the research are not even described as variables. Nevertheless, they seem to play similar role to traditional variables and therefore we believe it is useful to give these factors that label in order to enable comparisons throughout methods and disciplines.

As mentioned above, CL/CDA concentrates on discursive construction of groups, events or other social phenomena, another point of interest being more general features of discourse related to a subject. There are many social planes on which this research takes place: Almeida (2011) investigates the discursive construction of the Israeli-Palestinian conflict in U.S. newspapers, Hidalgo Tenorio (2011) is interested the way British press represented the death of Saddam Husain's sons and Weninger (2010) focuses on the construction of partnership in the redevelopment discourses. Features of discourse are included in the focus of MacDonald and Hunter (2013), who describe distinctive linguistic features used in documents relating to security for Olympics in London 2012 or of Alcaraz-Ariza (2002) in her presentation of linguistic-rhetorical strategies used to convey evaluation in medical book reviews. Therefore, both discursive constructions of chosen phenomenon and general features of discourse might be considered as most common dependent variables in the reviewed studies.

CL/CDA researchers also try to incorporate some context in their studies due to the crucial role of context in traditional qualitative CDA (Breeze, 2011). Incorporation of context into corpus-supported research happens mainly through including some independent variables such as:

- the source of the text. Gabrielatos, Baker (2008) compare the construction of immigrants in British broadsheets and tabloids;
- political views of the producer of the text. In analyzing parliamentary debates, Baker (2006) and Bachmann (2011) juxtapose discourses constructed by those in favor of the discussed bill with these constructed by the opponents of it;
- the language of the author. Mother tongue is considered as an important factor by Freake et al. (2010) when reconstructing the identity of people living in Québec.

## 3.5 Operational definitions

The basic notion of discourse is most commonly operationalized as a set of patterns emerging from collocations, concordances or keywords. The researchers reconstruct the discourses through search for regularities either in collections of lexical items (collocation, keywords) or short excerpts of text (concordances).

Collocation is defined as "more frequent than random co-occurrence" (Stubbs, 2001). In most studies collocates are generated for words somehow associated with the main subject in question. After obtaining the list of collocates the researchers either group them in order to describe regularities or look for some more advanced patterns such as semantic prosody (Louw, 1993) or semantic preference (Stubbs, 2001). The first approach leads Freake et al. (2010) to state that in French, Québec is constructed as historical community. Whereas analysis of semantic prosody allows Mautner (2007) to discover that the word *elderly* has a stronger negative prosody when it is used as noun than when it is used as adjective. The last procedure helps Kamasa (2013) to describe the discourses related to IVF as dehumanizing due to the fact that the words used to describe in vitro fertilization in the analyzed texts share semantic preference of inanimate objects.

The discourses are also operationalized as patterns emerging from the analysis of keywords, which are defined as "lexical items that show marked frequency in one text compared to another" (Scott, 2013). The corpus that the main corpus or group of texts is compared to is called a reference corpus and might have either a general or specialized character. The first approach is adapted by Weninger (2010) who describes facilitated agency in urban redevelopment reports by comparing them to FROWN (a general corpus of one million words of written American English) in order to calculate keywords. Specialized reference corpora are used, for example, by Subtirelu (2013): aiming at a description of ideologies related to language, he compares the speeches of proponents and opponents of multilingual voting materials.

The last and most common way of operationalizing discourse is identification of patterns in concordances. Baker (2006) defines a concordance as "a list of all of the occurrences of a particular search term in a corpus, presented within the context that they occur in" (p. 71). This context is rather short and contains usually between 10 and 30 words on either side of the search term. The search term, called also node word, is usually one of the following:

- a word chosen on the basis of researchers' knowledge and intuition related to the investigated subject. For example, in order to describe the ideological position of two different friendly-fire reports Albakry (2004) analysis the concordance for "salient terms of representation, agency, and stance markers" of his choice;
- a word chosen on the basis of lexical sources. Lischinsky (2011) uses FrameNet and WordNet to collect a list of words related to the notion of the crisis he is interested in;
- a keyword. Keywords generated as a result of a comparison of pro- and anti-sides in the same-sex marriage bill debate are used by Bachmann (2011) to describe discourses related to this subject;
- a collocate of words important for the subject. All collocates of the word Muslim are examined and divided into thematic categories with the help of

concordance analysis by Baker et al. (2013) in their pursuit to reconstruct "the ways that the press has constructed the word *Muslim*"(p. 256).

### 3.6 Methods

All the above mentioned instances of operationalization are based on methods allowing the researcher to automatically generate a list of lexical or textual items which are then manually analyzed. The automatic part is performed by concordancing software, usually AntConc (Anthony, 2013) or WordSmith Tool (Scott, 2013). The lists are generated on the basis of a simple text search, as in case of concordances, or statistical calculations (collocations, keywords). At this stage, the researchers decide on certain settings of the search, such as the length of the context that is presented in a concordance line, the statistical measure that is used to generate the list of collocations or keywords or the distance between the node word and its potential collocates. Manual analysis on the other hand is only rarely supported by any technical means; however, concordances are sometimes alphabetically sorted by words adjacent to the node word to facilitate the search for patterns (Baker, 2006), while collocates and keywords might be thematically grouped on the basis of automatic semantic tagging (Rayson, 2013).

### 3.7 Data

The collections of texts used in the studies mentioned above ranged in size from hundreds of thousands to hundreds of millions words. Just as in the case of general corpus linguistics, in corpus-supported CDA we may divide the corpora into two types: general corpora and specialized corpora. In the case of using general corpora, the analyses are more lexically oriented, for instance Orpin (2005) describes the similarities and differences in the usage of various words semantically related to corruption on the basis of the Bank of English. On the other hand, specialized corpora are used to describe discourses related to a social phenomenon in more detail. In most cases the corpora consist of press texts (e.g. Albakry, 2004; Alcaraz-Ariza, 2002; Almeida, 2011; Gabrielatos & Baker, 2008), official documents (e.g. Kamasa, 2013; Lischinsky, 2011; Subtirelu, 2013) or on-line texts (e.g. Lukac, 2011; Mautner, 2005; Prentice, 2010).

### 3.8 Sampling

This stage of research design is performed only by those researchers who wish to work on specialized corpora and have to face the problem of building their own corpus. They approach this problem in two main ways: either they decide to work on the whole body of available material or they develop some procedure for text sampling. In the first case (we might refer to it as population research, according to the social research tradition) the research problem is formulated in such a way that it is possible to analyze all texts relevant to the problem. This is the case for Edwards (2012) who is interested in diachronic changes in the self-presentation of British National Party in its manifestos. If the corpus needs to be sampled form wider collections of texts, it is most commonly accomplished on the basis of one of the following:

- rankings: Alcaraz-Ariza (2002) works with reviews published in the best medical journals;
- time of creation: Hidalgo Tenorio (2011) studies all news articles published in the UK one day after Saddam Hussein was captured by US troops;
- aboutness: Koller (2004) looks for war metaphors in articles describing businesswomen and businessmen;
- intuitive importance: Forchtner and Kolvraa (2012) base their study on speeches of politicians they consider to be important;

### 3.9 Presentation of results

Most frequently the results of a study are presented as a list of relevant collocations, keywords or concordances followed by a description of the main patterns observed on these lists. In some cases, when the study has a comparative character, statistical measures are used to show the quantitative significance of the observed differences. For instance Almeida (2011) points out that Palestinian and Israeli authorities are cited with different frequencies in the US newspaper she investigated. Another mode of results' presentation is used by Bachmann (2011): he starts with a description of the discourse and then uses the results (keywords and concordance lines) to justify his identification of the suggested discourse.

### 3.10 Interpretation of results

The interpretation of the results mainly takes two forms: descriptive or explanatory. In the first, the results are interpreted in the form of a description of discourses they construct: Lukac (2011) on the basis of a combination of quantitative and qualitative methods presents the results obtained in which she concludes that there is a discourse of anorexia as a close friend on the blogs of pro-ana community. When an explanatory factor is added to the interpretation, the role of the previously identified results in the relation of power and domination is presented as in the case of Bachmann (2011) who states: "Marriage is constructed as having an unquestioned appeal, being the model for relationships, against which all other forms are measured. One can argue that the introduction of a different concept for gay men and lesbians which mirrors marriage reinforces 'the dominance of heterosexuality as a normatively better way of life' (Richardson, 2004, p. 10)" (p. 101).

## 4. Conclusions

After nearly two decades of corpus-supported critical discourse analysis the individual studies might differ in the subject, the data they are based on or the choice of corpus techniques, but the overall approach to use corpus methods in CDA remains very similar. It has to be mentioned, however, that the presented overview is in no way exhaustive: neither have we analyzed all studies in the field nor have we presented all details concerning the research design. Nevertheless, we believe that we have presented main tendencies and most frequent methodological choices that could be observed in the studies analyzed. We also hope that the provided overview reinforces the usefulness of study design description as both a tool for planning one's own research and an analytic instrument appropriate for observing the main empirical tendencies in a research field.

## References

*Albakry, M.* (2004). U.S. "Friendly Fire" Bombing of Canadian Troops: Analysis of the Investigative Reports. *Critical Inquiry in Language Studies, 1*(3), 163–178.

*Alcaraz-Ariza, M. Á.* (2002). Evaluation in English-Medium Medical Book Reviews. *International Journal of English Studies*, (vol. 2 (1)), 137–153.

*Almeida, E. P.* (2011). Palestinian and Israeli Voices in Five Years of U.S. Newspaper Discourse. *International Journal of Communication*, (5), 1586–1605.

*Anthony, L.* (2013). AntConc: AntLab.

*Babbie, E. R.* (2010). *The practice of social research* (12th ed). Belmont, Calif: Wadsworth Cengage.

*Babiński, G.* (1997). Przygotowanie badań empirycznych [Prepearing empirical research]. In M. Malikowski & M. Niezgoda (Eds.), *Badania empiryczne w socjologii: wybór tekstów.* [Empirical research in sociology: a reader] *Volume 1* (pp. 300–313). Tyczyn: Wyższa Szkoła Społeczno-Gospodarcza.

*Bachmann, I.* (2011). Civil partnership – "gay marriage in all but name": a corpus-driven analysis of discourses of same-sex relationships in the UK Parliament. *Corpora*, (Vol. 6 (1)), 77–105.

*Baker, P.* (2006). *Using corpora in discourse analysis. Continuum discourse series.* London, New York: Continuum.

*Baker, P., Gabrielatos, C., KhosraviNik, M., Krzyzanowski, M., McEnery, T., & Wodak, R.* (2008). A useful methodological synergy? Combining critical discourse analysis and corpus linguistics to examine discourses of refugees and asylum seekers in the UK press. *Discourse & Society, 19*(3), 273–306.

*Baker, P., Gabrielatos, C., & McEnery, T.* (2013). Sketching Muslims: A Corpus Driven Analysis of Representations Around the Word 'Muslim' in the British Press 1998–2009. *Applied Linguistics, 34*(3), 255–278.

*Breeze, R.* (2011). Critical Discourse Analysis and Its Critics. *Pragmatics*, (21:4), 493–525.

*Burr, V.* (1995). *Introduction to social constructionism.* London: Routledge.

*Degano, C.* (2007). Dissociation and Presupposition in Discourse: A Corpus Study. *Argumentation, 21*(4), 361–378.

*Edwards, G. O.* (2012). A comparative discourse analysis of the construction of 'in-groups' in the 2005 and 2010 manifestos of the British National Party. *Discourse & Society, 23*(3), 245–258.

*Fairclough, N.* (2001). Critical Discourse Analysis as Method in Social Scientific Research. In R. Wodak & M. Meyer (Eds.), *Methods of critical discourse analysis* (pp. 121–129). London: SAGE.

*Forchtner, B., & Kolvraa, C.* (2012). Narrating a 'new Europe': From 'bitter past' to self-righteousness? *Discourse & Society, 23*(4), 377–400.

*Frankfort-Nachmias, C., Nachmias, D., & Hornowska, E.* (2001). *Metody badawcze w naukach społecznych.* Poznań: Zysk i S-ka.

*Freake, R., Gentil, G., & Sheyholislami, J.* (2010). A bilingual corpus-assisted discourse study of the construction of nationhood and belonging in Québec. *Discourse & Society, 22*(1), 21–47.

*Gabrielatos, C., & Baker, P.* (2008). Fleeing, Sneaking, Flooding: A Corpus Analysis of Discursive Constructions of Refugees and Asylum Seekers in the UK Press, 1996–2005. *Journal of English Linguistics, 36*(1), 5–38.

*Hidalgo Tenorio, E.* (2011). Critical Discourse Analysis, An overview. *Nordic Journal of English Studies*, (Vol. 10, No 1 (2011)), 184–210.

*Kamasa, V.* (2013). Naming "In Vitro Fertilization": Critical Discourse Analysis of the Polish Catholic Church's Official Documents. *Procedia – Social and Behavioral Sciences*, (95), 154–159.

*Kamasa, V.* (2014). The Polish Catholic Church on IVF. Corpus Assisted Critical Discourse Analysis. In A. Fodor, T. Eitler, J. Pethő-Szirmai, K. Vadai, & B. Molnár (Eds.), *Critical Approaches to Discourse Analysis Across Disciplines. Abstract book* (p. 116).

*Koller, V.* (2004). Businesswomen and war metaphors: 'Possessive, jealous and pugnacious'? *Journal of Sociolinguistics 8/1, 2004: 3^22*, (8/1), 3–22.

*Lischinsky, A.* (2011). In times of crisis: a corpus approach to the construction of the global financial crisis in annual reports. *Critical Discourse Studies, 8*(3), 153–168.

*Louw, B.* (1993). Irony in the Text or Insincerity in the Writer? The Diagnostic Potential of Semantic Prosodies. In M. Baker (Ed.), *Text and Technology* (pp. 157–176). Amsterdam: John Benjamins.

*Lukac, M.* (2011). Down to the bone: A corpus-based critical discourse analysis of pro-eating disorder blogs. *Jezikoslovlje*, (12.2), 187–209.

*MacDonald, M. N., & Hunter, D.* (2013). The discourse of Olympic security: London 2012. *Discourse & Society, 24*(1), 66–88.

*Mautner, G.* (2005). The Entrepreneurial University. *Critical Discourse Studies, 2*(2), 95–120.

*Mautner, G.* (2007). Mining large corpora for social information: The case of elderly. *Language in Society, 36*(01).

*Mohamad, Subakir Mohd Yasin, Bahiyah, A. H., Yuen, C. K., Zarina, O., & Azhar, J.* (2012). Linguistic Sexism In Qatari Primary Mathematics Textbooks. *GEMA Online™ Journal of Language Studies*, (Volume 12(1)), 53–68.

*O'Halloran, K.* (2009). Inferencing and cultural reproduction: a corpus-based critical discourse analysis. *Text & Talk – An Interdisciplinary Journal of Language, Discourse Communication Studies, 29*(1), 21–51.

*Orpin, D.* (2005). Corpus Linguistics and Critical Discourse Analysis: Examining the ideology of sleaze. *International Journal of Corpus Linguistics*, (10:1), 37–61.

*Prentice, S.* (2010). Using automated semantic tagging in Critical Discourse Analysis: A case study on Scottish independence from a Scottish nationalist perspective. *Discourse & Society, 21*(4), 405–437.

*Rayson, P.* (2013). Wmatrix: Lancaster University.

*Scopus* (2014). *Search results.* Retrieved September 08, 2014, from http://www.scopus.com.scopus0.bu-169.bu.amu.edu.pl/term/analyzer.url?sid=7BF2600A48CEEB26AF5DCCCD87F73222.Vdktg6RVtMfaQJ4pNTCQ%3a130&origin=resultslist&src=s&s=%28+KEY+%28+%22critical+discourse+analysis%22+%29+AND+TITLE-ABS-KEY+%28+corpus+OR+corpora+OR+%22corpus+linguistics%22+%29+%29&sort=plf-f&sdt=b&sot=b&sl=105&count=79&analyzeResults=Analyze+results&txGid=7BF2600A48CEEB26AF5DCCCD87F73222.Vdktg6RVtMfaQJ4pNTCQ%3a21.

*Scott, M.* (2013). WordSmith Tools. Liverpool: Lexical Analysis Software.

*Stubbs, M.* (2001). *Words and phrases: Corpus studies of lexical semantics.* Oxford [England], Malden, MA: Blackwell Publishers.

*Subtirelu, N. C.* (2013). 'English… it's part of our blood':: Ideologies of language and nation in United States Congressional discourse. *Journal of Sociolinguistics 37–65*, (17/1), 37–65.

*van Dijk, T. A.* (1993). Principles of critical discourse analysis. *Discourse & Society, 1993*(4(2)), 249–283.

*van Dijk, T. A.* (2001). Multidisciplinary CDA: a plea for diversity. In R. Wodak & M. Meyer (Eds.), *Methods of critical discourse analysis* (pp. 95–120). London: SAGE.

*Weninger, C.* (2010). The lexico-grammar of partnerships: corpus patterns of facilitated agency. *Text & Talk – An Interdisciplinary Journal of Language, Discourse & Communication Studies, 30*(5), 591–613.

*Wodak, R.* (2011). Critical linguistics and critical discourse analysis. In J. Zienkowski, J.-O. Östman, & J. Verschueren (Eds.), *Handbook of pragmatics highlights: v. 8. Discursive pragmatics* (pp. 50–69). Amsterdam, Philadelphia: John Benjamins Pub. Co.

# Part Three: Language Analysis

Eva Malková
*Masaryk University, Brno, Czech Republic*

# Ellipsis and sentence fragments in Ian McEwan's *Amsterdam*: their effect on meaning

**Abstract:** Ian McEwan's fiction offers numerous opportunities to observe patterns in characters' discourse. One of them may be a pattern which involves omitting one or more words which are usually parts of the grammatical structure – fragments or cases of ellipsis. We will take a look at examples from the text of McEwan's *Amsterdam* (1998) and at how can these fragments be interpreted as well as how they affect the way the characters' utterances are understood. The questions raised here are: does the fact that there is actually less than a full sentence make the conversation more difficult to comprehend? Does it violate Grice's maxim of quantity? What leads the author to including such pattern in the discourse so frequently?

Keywords: ellipsis/fragments, *Amsterdam*, discourse, characters, meaning.

## 1. Introduction

Works of fiction and the discourse of the characters included in them are frequently full of various observable patterns. Each author has a unique style which is usually reflected in the text and subsequently in the characters' conversation. What is said by them depends a lot on the author's decisions and therefore differs more or less from an authentic conversation. However, there are means and tools that can be used to make the conversation among the characters as realistic as possible while preserving the tension that keeps the reader interested in the story till the very last page.

In Ian McEwan's novel *Amsterdam* (1998), one of the very frequent patterns is inclusion of sentence fragments or ellipsis (omission of sentence structure parts) within his characters' discourse. Some cases of omission of sentence structure parts cannot exactly be defined as ellipsis (after all, there are types of sentences which only consist of one or two constituents, for example). The preceding context (whether linguistic or situational) is also of significant importance. We are therefore adopting John Merchant's (2005) term "*fragments*" which entails structures with no linguistic antecedent (p. 1).

## 2. Review of literature

Apart from Ian McEwan's *Amsterdam* (as the text subject to analysis), the literature used as theoretical framework for the paper includes M.A.K. Halliday's and Christian Matthiessen's *An Introduction to Functional Grammar* (2005) and the *International Encyclopedia of Linguistics* (2003), as well as John Merchant's article *Fragments and ellipsis* (2004), which provide a classification of the ellipsis and sentence fragments phenomenon. These are supported and complemented by Marjorie McShane's *A theory of ellipsis* (2005) and, most importantly, Michael Swan's *Practical English Usage* (2003), which provides detailed grammatical categorization of ellipsis. Roger Fowler's *Linguistic Criticism* (1996), Tim Wharton's *Pragmatics and Non-Verbal Communication* (2009), John Lyons' *Semantics, Vol 2.* (1977) and M. A. K. Halliday's *Spoken and Written Language* (1989) have provided background for the general hypothesis.

## 3. Methodology

As for the basic outline of the plot of the novel, *Amsterdam* is a story focused on two middle-aged men, Clive Linley (a world renowned composer) and Vernon Halliday (the editor of *The Judge* magazine) who are best friends and have a lot in common, including the woman they both used to be in love with – Molly Lane. Both men struggle in a way with their lives, facing every-day problems (especially moral dilemmas – potential publishing of scandalous photographs in Vernon's case, and (not) helping a woman who is being raped in Clive's) and consequences that their decisions have on them and other characters further in the story. Their intense relationship even leads to arguing and fighting with each other. Despite their differences they remain able to overcome their conflict and preserve their friendship till the very end. Molly, the woman they both loved, is currently dead, but she still operates as the factor that unites not only these two men, but also other people involved in the story. Having watched Molly deteriorate and expire in what they consider an undignified way, the two men promise each other to end the other one's life before they reach such state. The story then inevitably culminates in both of their suicide, or, more fittingly, a mutual murder.

Let us focus on the conversation among the characters and its specifics. The author is very good at involving the reader in the story as actively as possible by letting the characters explicitly say only certain things, leaving the rest to only get revealed gradually. McEwan's characters are often very brief and of few words. As has already been mentioned above, if we were looking for a characteristic trait of the characters' discourse in the novel, we would find out that it often includes short utterances or comments which drop one or more words which are usually parts of the grammatical structure of the English sentence (Subject, Verb, Object). They are, in fact, fragments, but that does not necessarily mean that they are not understandable.

> "A conversation consisting entirely of grammatically complete text-sentences would generally be unacceptable as a text; and it is part of the language-competence of a speaker of the language (if not of his linguistic competence in the narrower sense) that he should be able to produce grammatically incomplete, but contextually appropriate and interpretable, sentence fragments." (Lyons, p. 589).

Except for what we (according to John Merchant (2004)) call *fragments*, McEwan of course also includes actual ellipsis in his characters' discourse. Ellipsis can be classified in various ways; the *International Encyclopedia of Linguistics* defines it as "… sentences containing gaps that are interpreted under identity to some other constituent, not necessarily in the same sentence"(p. 513) and categorizes it into surface structure ellipsis (which can be embedded) and gapping (which can only occur in coordinate structures). It is also important to mention M.A.K. Halliday's (2004) categorization into anaphoric and, more significantly, exophoric ellipsis: "In this type of ellipsis the clause is not presupposing anything from what has gone before, but simply taking advantage of the rhetorical structure of the situation, specifically the roles of the speaker and listener. […] Such clauses have, in fact, a thematic structure; but it consists of Rheme only. The Theme is (part of) what is omitted in the ellipsis." (p. 111) Michael Swan's *Practical English usage* textbook has proven to be helpful in categorizing ellipsis depending on which part of the sentence is omitted.

My hypothesis is based on the assumption that spoken and written language (as Halliday (1989) says) are organized differently, as well as discourse amongst various fictional characters varies, especially in the degree of stylization. Fictional discourse, therefore, could be expected not to employ the device of ellipsis and fragmenting so frequently, however, McEwan's characters include them to an extensive amount.

Therefore the questions raised are whether this strategy is not counterproductive – does such richness in ellipsis in any way affects the meaning, or rather, understanding of the utterances among the characters? Does it violate Grice's maxim of quantity? And if not, what are the author's potential reasons for such speakers' choices?

Real people's daily conversation is commonly full of ellipsis occurrences, we use it frequently and are able to infer from the context, so the assumption is that the author wants to make his characters' conversation as similar to authentic conversation as possible.

## 4. Analysis

To understand ellipsis correctly it is necessary to be able to recover the omitted material (usually from appropriate antecedent), as can be seen in many of the following examples. *Amsterdam* mostly includes cases of surface ellipsis. There is no problem understanding here at all, on the contrary, we could say it is even a matter of economy – less effort on the side of the speaker without requiring much effort on the side of the hearer. "The linguistically encoded logical form which is the output of the mental grammar is simply a starting point for rich inferential processes guided by the expectation that speakers will conform to certain standards of communication." (Tim Wharton, p. 9)

Most of the occurrences of ellipsis in *Amsterdam* are cases of verb phrase deletions. The following excerpt is a brief exchange between Clive and Vernon about the deceased Molly and her disease:

> *"'She never knew what hit her.'*
> *'When she **did** it was too late.'" (Amsterdam, p. 3)*

Instead of repeating what Clive said, Vernon logically uses just the auxiliary verb. Cases of gapping are not as frequent, but not non-existent.

The omitted part is highlighted in bold. In both examples the referent is syntactically accessible. As Swan's textbook states, in co-ordinate structures (like this one) it is not uncommon to omit parts which otherwise only would have been unnecessary repetitions (p. 172). Ellipsis of this kind is frequently used in whatever type of communication – it reduces the effort on both sides (hearer and speaker).

### 4.1 NPs, nouns, verbless structures

In general, McEwan is especially fond of using very brief sentences – vast majority of them are sentences consisting merely of nouns or noun phrases. Let us take a look at the examples of the above mentioned *fragments:* it is a conversation between two men who know of each other by reputation, but have never formally met before and this is their first time talking to each other. Clive Linley is talking to Julian Garmony (the Foreign Secretary) at Molly's funeral. Garmony is Molly's former lover (just like Clive and Vernon), so it is not surprising that Clive does not find the conversation pleasant at all. Garmony tells Clive that his wife knows a few of his piano works by heart, to which Clive replies that she must indeed be good.

> *"Garmony was looking around now, securing his audience. 'She was brilliant.* ***Goldsmith's, then Guildhall. A fabulous career ahead of her****… 'He paused for a comic effect. 'Then she met me and chose medicine.'" (Amsterdam, p. 14)*

Julian Garmony uses no predicates when mentioning his wife's musical achievements. Each of his utterances only contains a subject: "Goldsmith's" and "Guildhall" can be traced to no syntactic antecedent, but both are places very well-known. This, therefore, is only a fragment – the fact that Clive is a composer and that they had been talking about his music before created enough context. To drop (not only) the verb phrase but also other parts of the full sentence here undoubtedly means that the Foreign Secretary wanted to stress his wife's playing brilliance and then impress Clive even further in the same manner by dropping the verb phrase also from the following sentence. *Goldsmith's*, *Guildhall* and *fabulous career* are the bombs that he is throwing on him with no need of further specification (such as verb phrase would contain). Michael Swan's textbook includes an ellipsis category of "well-known names" (p. 184), the last words of which are often dropped – in this case they correspond to what Garmony omits when he mentions the institutions that represent his wife's musical achievements. They could be categorized under Marjorie McShane's (2005) semantic ellipsis, "… the non-expression of information that, although syntactically not required, is necessary for a full semantic representation of the sentence." (p. 25), which could make it possible to classify this particular excerpt as actual ellipsis, not a fragment.

Let us move on to another example: Molly's husband George offers Vernon and his magazine scandalous photographs of Julian Garmony, the Foreign Secretary. They are about to be published in the magazine and one of Vernon's subordinates, Frank, is looking at them in disbelief.

> *"But three days later, when Vernon was beginning to run the corridors, startled by the frenzy opposition and starting – but just slightly – to waver, he returned with Dibben to the same pub, to exactly the same booth, and showed him the photographs. Frank gazed at each one at length, without comment, simply shaking his head. Then he put them back in the envelope and said quietly,* **'Incredible. The hypocrisy of the man.'**" *(Amsterdam, p. 107)*

Frank practically only uses an NP and a mere adjective here, but there is no doubt that the hearer has no problem understanding. We could say that what Frank says here is almost close to an interjection – he expresses his irritation. That and the supposed feeling of frustration is why Frank limits himself to a fragment, or, a very brief verbless comment. "The hypocrisy" itself is a very negative word and also a very strong one to use about a person's character. As we can see, the negative part of the utterance is separated from the rest (more on the separation of negation below). It seems the author lets Frank drop the less informative parts and use nothing else but the negative NP – again, in order to emphasize how strongly he feels.

## 4.2 Negation separation

Moving on to more interesting samples from the text, there is a part of conversation between Vernon, the editor of *The Judge*, and Molly's husband, George, about the previously mentioned photographs of Garmony.

> *"They sat almost twenty feet apart, with the hissing fireplace between them. He had been alone for half a minute, Vernon thought, he might have crawled over to the fender and knocked the right side of his head on it. Even in company now, he did not feel right. 'I've seen the ABC figures,' George said gravely. '**Not good**.'"* (*Amsterdam, p. 52)*

This omission of both the subject and the predicate could be classified as exophoric ellipsis; George's last utterance is purely rhematic. Although these parts of the sentence are omitted, there's no confusion or misunderstanding or even violation of the maxim. The subject can be traced to George's previous utterance, the predicate is omitted entirely, but that has no negative influence on the overall meaning or understanding or the maxim. The interesting fact is that George separates the negative part of the sentence form the rest. Obviously, he wants to stress the part that carries the negation, and thus create enough tension. When the fact that Vernon is a bit hesitant is taken into consideration, it becomes evident that the main purpose of George's choice to separate the negation from the rest of the utterance is to make Vernon realize that he really needs his help. (The scandal in question concerns the Foreign Secretary – when Molly died, her husband came into possession of some photographs of him that she had made and that could have compromised his career if published).

Another example of a similar type of construction including an exophoric ellipsis comes from the above mentioned dialogue between Clive and Julian Garmony. The Foreign Secretary congratulates Clive on his commission for the Millennial Symphony:

> *"'Some of my colleagues wanted this pop star chap, the ex-Beatle. Anyway, how is it coming along. **Almost done?**'*
> *'**Almost**.'" (Amsterdam, p. 14)*

The Foreign Secretary's question is already elliptical; Clive's answer even further so. This example shows another verb phrase deletion, the referent being traceable to the previous sentence. It is Clive's speaker's choice what is interesting here – he could have just answered *"Yes"*, or *"Yes, it is"*, but instead he decides to repeat (part of) what the Foreign Secretary said – in order to stress that the answer is positive, however, only to a certain limit. This also creates a substantial amount of tension and suspense – by choosing to structure his answer this way, Clive conveys towards Garmony (who, as Foreign Secretary, is a figure of authority) an expression of undeniable defiance.

## 4.3 On ellipsis as a sign of intimacy

It may seem (and very frequently it is so) that ellipsis (as Roger Fowler says in his Linguistic criticism (1996)) is mostly emotional, suggests intimacy and therefore it is typical for people (or characters) who are generally more or less close to each other and know each other well, as the following excerpt (part of which we have already seen above) shows. It comes from a dialogue between Clive and Vernon.

> *"'You know, I should have married her. When she started to go under I would have killed her with a pillow or something and saved her from everyone's pity.'*
> *Vernon was laughing as he steered his friend away from the Garden of Remembrance. '**Easily said.** I can just see your writing exercise-yard anthems for the cons, like what's her name, the suffragette.'" (Amsterdam, p. 8)*

When replying to Clive, Vernon omits the subject (another example of exophoric ellipsis) which would have stood for Clive's entire utterance – there is no need to include it, after all, he is talking to his best friend and the conversation is informal. Ellipsis here then really implies closeness. However, this is not always the case. The identity of the discourse participants and their mutual relationships undoubtedly do influence their speakers' choices in the decision-making process within the conversation. On the other hand, omission of certain parts does not always signify closeness between the discourse participants. When we take a look at another sample from the novel, it is the police talking to Clive Linley on the phone, telling him that they will need him to help them identify a suspect in the case of the Lakeland rapist (whom Clive had seen before).

> *"The bell again, then silence. They'd gone. For a moment, the slender idea he had was lost. Then he had it, or part of it, and was just drawing the stem of a chord when the phone rang. He should have turned it off. In his irritation, he snatched it up. 'Mr Linley?'*
> *'Yes.'*
> *'Police. C.I.D. **Standing outside your front door. Appreciate a word.**'*
> *'Oh. Look, can you come back in half an hour?'*
> ***"Fraid not. Got a few questions for you. Might have to ask you to attend a couple of identity parades in Manchester. Help us nail a suspect. Shouldn't take up more than a couple of days of your time.** So, if you wouldn't mind opening up, Mr Linley...' (Amsterdam, p. 144)*

The policeman on the phone omits a great amount of sentence parts – articles, pronouns, verbs and verb phrases... His utterance only contains the parts necessary to understand it. This does not mean he knows Clive or that he wants to have an informal small talk with him, but (since it is a telephone conversation) rather, he wishes to make his contribution to the dialogue as brief and factual and probably also as understandable as possible, which means that understanding is hardly an issue here. Omission here enables him to avoid tiresome and potentially disturbing repetition. The words that the police officer uses are those that are supposed to bear the most stress; so for the sake of efficiency, he only utters those and leaves out all the rest. That may indicate that ellipsis can be a tool of implying relationships between characters, but it is certainly not exclusively a sign of intimacy or closeness. Interestingly enough, while the police officer only uses fragmented sentences when asking Clive out, the ending of his part of the dialogue is different. Although unfinished, the structure of his last sentence is not fragmented, and it expresses a degree of politeness. It would be therefore more fitting to say that words are omitted when the hearer is able to reconstruct the whole meaning without them; whether or not a closer relationship with the speaker exists. Ellipsis and fragmenting, however, can be a useful tool of expressing it.

## 5. Conclusions

To conclude what has been observed in the individual cases and examples from the text of the novel, ellipses in here hardly ever or never cause any ambiguities or problems in understanding at all, nor do they require more effort to do so by the hearer – actually it proves to be quite the contrary. There is enough information, which is necessary for unproblematic, full comprehension – i.e. the parts that are omitted from the sentence are not vital. Grice's maxim of quantity is meticulously fulfilled.

From what the excerpts from the text of the novel have shown, ellipsis and fragments may have several functions in Ian McEwan's fiction: they can be understood as a tool for letting the characters **stress** what they think is the most important part (by separating it from the rest of the utterance). As expected, they also have a role in making the characters' discourse more similar to **authentic conversation** among people (including the inferential process which enables the speaker to spend less energy while causing no difficulty understanding on the side of the hearer). This leads to further assumptions that they can as well in some cases be a tool of **indicating the intensity of relationships** between the characters. Quite frequently, they can contribute to the elements **creating tension and suspense** in the text.

## References

### Books

*Fowler, Roger.* (1996). *Linguistic criticism.* Oxford University Press.

*Frawley, William J. (Ed.).*(2003). *International encyclopedia of linguistics.* Oxford University Press.

*Halliday, M.A.K., Matthiessen, Christian.* (2004). *An introduction to functional grammar* (3rd Edition). London: Arnold.

*Halliday, M.A.K.* (1989). *Spoken and written language.* Oxford University Press.

*Lyons, John.* (1977). *Semantics, vol. 2.* New York: Cambridge University Press.

*McShane, Marjorie J.* (2005). *A theory of ellipsis.* New York: Oxford University Press.

*Swan, Michael.* (2003). *Practical English usage.* Oxford University Press.

*Wharton, Tim.* (2009). *Pragmatics and non-verbal communication.* New York: Cambridge University Press.

### Articles

Merchant, John. (2003). Fragments and ellipsis. *Linguistics and Philosophy,* 27.6:661–738. Retrieved September 3, 2014, from <*home.uchicago.edu/merchant/pubs/fragments.pdf*>.

Olga Berdina
*Mariupol State University, Ukraine*

# Meaning change of gradual verbs denoting colour in English

**Abstract:** The article deals with meaning change of gradual verbs denoting colour – that is the forming of gradual seme in the semantics of the English verbs. Gradation is considered to be an important element of measure and degree. The aim of the article is confirm that gradation is a linguistic category, and to distinguish the nature and the results of the meaning of change as far as gradual verbs of colour in English is concerned. The changes of gradual meaning are realised by means of nuclear, differential, potential semes and a transfer (metaphorical change). Along with this broadening and narrowing, amelioration and pejoration of gradual meaning in the semantics of colour verbs are semantic changes have been determined.

Keywords: gradual verbs, gradation, meaning change, broadening, narrowing, pejoration, amelioration.

## 1. Gradation as linguistic category

Our research focuses on the peculiarities of evolution of gradual meaning in the semantics of verbs denoting colour in English. The *aim* of this article is to confirm that gradation is a linguistic category; to distinguish the nature and the results of the meaning of the changes as far as gradual verbs of colour in English is concerned. The meaning of verbs denoting colour can be complicated with the gradual seme. Such verbs as *engolden, dilute, raddle* can be referred to gradual verbs.

Interest in gradation as a linguistic phenomenon is connected first of all with the ability of gradation to express reality in its dynamics, constant changes and development. The theoretical grounds of this article are illustrated by the analysis of different approaches to the study of gradation in linguistics (Sheygal, 1991; Kolesnikova, 1998, 1999, 2010; Khalina, 1993; Marchuk, 2008; Sepir, 1944/1985; Bolinger, 1972; Bierwisch, 1989; Osgood, 1957; J. Rasmussen, 1996; Kennedy, McNally, 2005) and change of semantic meaning (G. Stern, 1931; L. Bloomfield, 1933; S. Ullmann, 1931/1972; A. Blank, 1998; E. Coseriu (1973/2010), D. Shmelev (1973/2008), Geeraerts, 1983).

## 1.1

The main concepts of gradation study as language category were written in the article "Gradience. A Study in Semantics" by E. Sepir (1944/1985). The author educed gradation as a linguistic category, outlined promising directions of its studying and distinguished the main terms of this category (Sepir, 1944/1985). According to *E. Sepir's* concept, units having the qualitative component in their semantic structure or ability to the process of comparison can be called gradual (Sepir, 1944/1985). The fundamental work written by D. Bolinger (1972) "Degree Words" has become continuation of the main *E. Sepir's* idea. The author has studied semantic peculiarities of intensifiers and their combination with other parts of speech (Bolinger, 1972). G. Bolinger stressed that "many verbs and nouns are regularly intensified; the explicit treatment concerns nouns and verbs mainly" (1972, p. 17). M. Bierwisch (1989) distinguished the way gradation is expressed in linguistics.

According to the concept of H. Sheygal (1991), we understand the gradation of the action as the continual growing or lowering of the present feature on the gradual scale. Gradation is connected with the qualitative estimation and at the same time it is a universal type of semantic relationships in the language (Sheygal, 1991, p. 83). In our research ***gradation*** is understood as a linguistic category which is treated as special semantic relationship in language when some quality of object can be changed or developed according to the particular gradual scale and norm.

M. Bondarenko (2001) has made an interesting review as far as the problem of gradation of the works written by Russian, Ukrainian and English scholars is concerned. The author claims that gradation is the complex category which contains such categories as intensity, comparison, and connotative component (Bondarenko, 2001). L. Marchuk (2008) points out that the indicating of gradation as pragmatic is dictated with its universal ability of communicative needs. The author confirms that gradation "is the means of knowledge about measure, grade and intensity of different phenomena" (Marchuk, 2008, p. 12).

While agreeing with the basic concept of E. Sapir (1944/1985) and the functional approach to the study of language units O. Bondarko (1984), S. Kolesnikova (1999) understand gradation as a lexical-semantic category (p. 13). The authors admit that most of the words can enter into gradual relationship and express gradability of the meaning (Kolesnikova, 1999).

In addition, in linguistics separate groups of gradual vocabulary are studied: characteristics of nouns (Kolesnikova, 1999; Fedotova, 2005; Sheygal, 1991) adjectives (Galich, 1981; Korzhik, 1999; Kostusyak, 2002; Kotnyuk, 1986; Mezenina, 1991; Pasichnik, 1998), verbs (Isaeva, 2009; Kolesnikova, 2010; Sheygal, 1991), adverbs (Nazarova, 2009), the structural parts of speech (Zastrovskiy, 2006; Kireeva, 2010).

## 2. The nature and the results of the meaning change of gradual verbs

Word-meaning is liable to change in the course of the historical development of language. The development and the change of the semantic structure of a word is always a source of qualitative and quantitative development of the vocabulary. All the types discussed depend upon some comparison between the earlier (whether extinct or still in use) and the new meaning of the given word (Ginzburg, 1979). Language must continuously change for that, in order to function. It is created with the help of shifts and ceases to exist, when stops the development (Coseriu, 1973/2010, pp. 204–205). Certain changes which result in appearing of new and disappearing of others meanings happen in the meanings of words. "They are caused by certain variations in the objective reality and by the peculiarities of internal semantic relationship" (Shmelev, 1973/2008, p. 19).

### 2.1 Scholars' classifications of semantic change

Many scholars have attempted to classify changes in meanings but few have written the processes of semantic changes.

Recent overviews have been presented by A. Blank (1999). One of the first major works were M. Bréal (1899), G. Stern (1931), L. Bloomfield (1933) and S. Ullmann (1962). Studies beyond the analysis of single words started with the word-field analyses of J. Trier (1931), who claimed that every semantic change of a word would also affect all other words in a lexical field. His approach was later refined by E. Coseriu (1973/2010). As stated above, the most currently used typologies are those by L. Bloomfield (1933) and A. Blank (1998) and other typologies are listed below. A number of classification schemes have been suggested for semantic change. The most widely accepted scheme is by L. Bloomfield (1933). The author distinguished narrowing, widening, metaphor, metonymy, synecdoche, litotes, hyperbole, degeneration, elevation (Bloomfield, 1933, p. 121).

However, the categorization of A. Blank (1998) has gained increasing acceptance: metaphor, metonymy, synecdoche, specialization of meaning, generalization of meaning, cohyponymic transfer, antiphrasis, auto-antonymy, auto-converse, ellipsis, folk-etymology. The author considers change of meaning classification problematic, though, to include *amelioration* and *pejoration* of meaning as well as strengthening and weakening of meaning. According to A. Blank (1998), these are not objectively classifiable phenomena.

S. Ullmann (1972) distinguishes between nature and consequences of semantic change – to the nature of semantic change he referred metaphor, metonymy, folk-etymology, ellipsis; to the consequences of semantic change – widening of meaning: raise of quantity, narrowing of meaning: loss of quantity, amelioration of meaning: raise of quantity, pejoration of meaning: loss of quality (p. 42).

G. Stern (1931) tried to classify the change of meaning and distinguished such types as: 'substitution', 'analogy', 'shortening', 'nomination', 'transfer', 'permutation' and 'adequation' (pp. 166–168). M. Breal (1899) was probably the first to emphasize the fact that in passing from general usage into some special sphere of communication a word, as a rule, undergoes some sort of specialisation of its meaning. For example, it can be illustrated by the word '*degrade*' which has the first meaning '*reduce to lower rank, depose from a position of honour*', but later the new specific meaning with gradual component was observed: *spec. reduce or tone down in colour*. It should be pointed out, the general (not specialized) meaning is also very frequent in present-day English.

## 2.2

D. Shmelev (1973/2008) emphasizes that "meaning of polysemantic word combines into semantic unity through semantic relationships that exist between them on a common semantic associations (metaphor, metonymy, functional community)" (pp. 86–87). The same opinion was expressed by R. Ginzburg, S. Khidekel, G. Knyazeva and A. Sankin (1979), who believe that the changes in the meaning of words can occur in two ways: *metaphor* or *metonymy*. "Similarity of meanings or metaphor may be described as a semantic process of associating two referents – one of which in some way resembles the other" (Ginzburg et al., 1979, p. 30). It can be demonstrated by the word 'green' which in Old English had the meaning '*become covered (over) with verdure*' but in Early twentieth it was changed and got into '***fig.*** *rejuvenate*' on the base of the similarity of verdure, always green and the age or time. Contiguity of meanings or metonymy may be described as the semantic process of associating two referents one of which makes part of the other or is closely connected with it. The author proved this by the use of the word *toungue* when '*the organ of speech*' in the meaning of '*language*' (Ginzburg et al., 1979, p. 30). As the result of such changes, the meaning of the word can be broaden or narrow. Agreeing with the concept of R. Ginzburg and S. Khidekel (1979), we distinguish two main groups of changes in the semantic of the words, as a rule accompanied by a change in the denotational' component: a) *pejorative development* or the acquisition by the word of some derogatory emotive charge, and b) *ameliorative development* or the improvement of the connotational component of meaning.

### 2.3

First of all, semantic changes occur in the group of words (Shmelev, 1973/2008, p. 19). That is the research of semantic changes in verbs with gradual meaning is possible within the lexical-semantic group.

The problem of semantic classification of verbs belongs to the most popular issues of nowadays. Its complexity and diversity is seen in the variety of research approaches to the study of verbal meaning and of course in diversity of semantic classification of verbs. According to Vinogradov (1947) "the verb is the most complicate and sizable grammatical category" (p. 35).

## 3. Empirical material: characteristics

### 3.1

The body of linguistic material makes 75 colouring verbs with gradual semantics carried out by continuous sampling of the English Language Dictionary "The Shorter Oxford English Dictionary on Historical Principles" in two volumes, edited by William Trumble and Angus Stevenson [SOED].

A number of methods of linguistic analysis were used. The componential analysis has made it possible to distinguish gradual meaning of verbs denoting colour. On the basis of diachronic analysis the character of evolution of gradual meaning has been traced. The quantitative analysis allowed finding out the productivity of change of gradual meaning.

### 3.2

To select the verbs of gradual semantics from the general fund language has been established a number of explicit the gradual component in the lexical definition of the verb. For corpus forming the special words-graduators were used. They are such verbs which indicate the degree of manifestation of the feature in the definition of the word, such as *too, such, so, like, as, quickly, sharply, more* and others. So the qualitative component occurs in the lexical meaning of the verb which creates the conditions for gradation: heighten – make (a colour) more luminous or more intense; lit – blush deeply; marble – stain or colour (paper, the edge of a book, etc.) to look like variegated marble.

There should obviously be words-identifiers in the definition of gradual verbs which help to select the verbs with colour seme. Such words prove the attachment to the right lexical-semantic group: *stain, colour, shade, tinge, blush, redden, blacken, whiten, make + pearly, red, black, etc.*

## 4. Evolution of colour verbs with gradual semantics

### 4.1

The corpus of verbs denoting colour with gradual component in English is represented by 75 verbs. Evolution of gradual meaning is treated here as the process of old semantic extension resulting in complete replacement of the old meaning by a new one or when the initial meaning is completely lost and replaced by a new one. For instance, initially the gradual meaning '*stain or paint a blood-red colour*' had been developed in semantics of the verb *sanguine*. But later it evolved to another gradual meaning '*stain with blood*' which went out of use.

### 4.2

The results of semantic change can be generally observed in the changes of the denotative meaning of the word (narrowing and broadening of meaning) or in the alteration of its connotative component (amelioration and pejoration of meaning).

#### *4.2.1*

Under **semantic broadening** of the gradual meaning we understand the process in which the meaning of a word becomes more general or more inclusive than its historically earlier form. Group of gradual meaning which develops is the most numerous (51 %). Gradual meaning '*become or grow white or whiter; turn pale, esp. from fear*' of the word '*whiten*' demonstrates the example of semantic broadening. The first meanings of this word were '*make white or whiter*'; '*make pale or paler*'; '*spec. (a) whitewash; (b) bleach*' within the time the meaning of '*whiten*' was developed. The extension of the word meaning can be illustrated by the word *smut* which originally meant '*stain with some fault or imperfection; defile, corrupt*' but later '*blacken, smudge*' where the last meaning of the verb is gradual. It should be pointed out, the semantic broadening of gradual meaning obviously happened in the period of the Early New English (43 %).

#### *4.2.2*

Change in the denotative meaning of the word may result in **narrowing of gradual meaning.** This may be illustrated by the semantic development of the word '*ruddy*' which used to denote '*ruddiness, esp. in facial complexion*' as a noun in Middle English but in Late seventeenth century denotes only '*make or become ruddy; redden*' where the last meaning is gradual. This is also the case with the word '*cork*' which in old English denoted '*provide with a cork as a sole,* *float, etc'*, but in Modern English denotes *'blacken with burnt cork'*. This semantic change of gradual meaning is not numerous (44 %). There are some cases of specialization of gradual meaning (3 % from the whole amount of meaning narrowing). For example, we can observe specialization of meaning in the case of the verb *'heighten'* which had the meaning *'exalt; elate, excite; beautify'* and has now acquired a restricted and specialized meaning *'spec. make (a colour) more luminous or more intense'*.

*4.2.3*

Less productive group contains the verbs with the gradual meaning which have been ousted from use, i.e. **obsolete** meanings (5 %). This may be illustrated by the semantic development of the word *'black'* which used to denote *'be or become black'* from Middle English till the Middle nineteenth century. Disappearing of gradual meaning can be illustrated as well by the word *'gild'* which used to denote *'impart a brilliant colour or flush to (the face)'* only in the Early seventeenth century.

*4.2.4*

The analysis of the verbs with the gradual meaning has proved the changes in the connotative meaning: **amelioration** (20 %) and **pejoration** (31 %). The ameliorative development of the connotational meaning may be observed in the change of the semantic structure of the word *'green'* which in one of its meanings originally denoted *'become green'*, but now – *'make (an urban area) more verdant by tree-planting'*. The case of pejorative development of the gradual meaning can be illustrated by the word *'smutch'* which had its first meaning *'to soil, stain, blacken'* but later came to denote gradual meaning *'blacken, make dirty, smudge'*. It should be stressed that changes in connotational meaning are isolated cases. Almost a half gradual meanings don't have any changes in their connotational part (49 %), for example: *engolden – make or become golden.*

**4.3**

Speaking about the results of the gradual meaning changes, we should mention the nature of these changes. The gradual meaning can be based on the nuclear seme of the word. This case can be illustrated by the word *'black'* when the gradual meaning *'make black'* is based on the nuclear seme 'black' in the first meaning *'be or become black'*. Such cases make 40 % from the whole corpus of gradual colour verbs.

36 % of gradual meanings are based on the differential semes. It can be clarified by the word *'crimson'* when the gradual meaning *'become crimson, blush'* is

based on the differential seme *'crimson'* of the first *'make crimson'*. Less numerous is the group where the gradual meaning is grounded on potential seme (16 %), e.g.: *fade – 1) lose strength or vitality // grow weak, waste away; 3) lose brightness or brilliance of colour // grow dim or pale*. The example proves that semes *brightness or brilliance* are not observed in the initial definition of the word *'fade'* but there is the association between the man's strength and energy and lose of brightness by colour.

One of the most striking types of the nature by metaphor of semantic change is triggered by a **metaphor**, a figure of speech based on a perceived similarity between distinct objects or actions. It has been observed in the words *'yellow'* acquired in Old English *'fade to a yellow colour'* because of similarity of yellow shade to the degradation of something or losing previous features (**fig.**: "All that is left is a batch of *yellowing declarations* of good intentions" (Guardian Weekly)). Metaphorical ground for the gradual meaning is rare in English verbs of colour (8 %).

## 5. Conclusion

### 5.1

Category of gradation the linguistic category that exists on the lexical-semantic level and is treated as special semantic relationships in language when some quality of the object can be changed or developed according to the particular gradual scale and norm. Semantic features of gradation are characterized by measuring, relativity and dynamics.

### 5.2

The meaning of the word is changed in the course of the historical language development. The analysis of the specific features of the gradual verbs have made it possible to distinguish the nature and the results of meaning change of verbs with gradual semantics in English. Change of meaning is effected through association between the existing meaning and the new. This association is generally based on the similarity of meaning (metaphor), on the nuclear, differential or potential semes. The most representative are the verbs of colour with the gradual meaning based on the nuclear semes (40 %). Mostly the gradual meaning comes from the first lexical variant of meaning.

### 5.3

Semantic changes in the denotational component may bring about the broadening or the narrowing of meaning. The gradual meaning can be found as the result of broadening of the colour verb meaning. This result of change of meaning is the most representative among others (51 %) and is common for the Early New English period. In the discussion of semantic changes we distinguished among broadening the cases of generalization (2 % from the whole amount of meaning broadening). The narrowing of the meaning is less productive (44 %) than broadening and is typical for the Middle English period. Also we can observe specialization of gradual meaning (3 % from the whole amount of meaning narrowing). The cases of obsolete meanings are isolated and make only 8 %. The change in the connotational component may result in the pejorative or ameliorative development of meaning but these cases are isolated.

Therefore the lexical-semantic group of the verbs denoting colour with the gradual component is not representative, though the regular change of meaning has been observed. The aim of our further research can be: peculiarities of semantic changes in other groups of verbs with the gradual meaning illustrating their nature, results and possible objective reasons.

## References

*Apresyan Yu.D.* (1995). Лексическая семантика. Синонимические средства языка. Москва: Школа «Языки русской культуры».

*Bierwisch M.* (1989). The semantics of gradation. In dimensional adjectives. Berlin: Springer-Verlag.

*Blank, A.* (1999). Why do new meanings occur? A cognitive typology of the motivations for lexical semantic change. In A. Blank & P. Koch (Eds.), *Historical Semantics and Cognition* (pp. 61–89). Berlin & New York: Mouton de Gruyter.

*Blank, A.* (2003b). Words and concepts in time: towards diachronic cognitive onomasiology. In R. Eckardt & K. v. Heusinger & Chr. Schwarze (Eds.), *Words in Time. Diachronic Semantics* (pp. 37–65), Berlin & New York: Mouton de Gruyter.

*Bloomfield, L.* (1933). *Language*. New York: Allen & Unwin.

*Bolinger D.* (1972). *Degree words.* Paris: Mounton.

*Coseriu E.* (2010). Sincronia, diacronia e historia (I.A. Melchuk,Trans.). Madrid. (Original work published 1973).

*Douglas H.* (2001). Etymological Dictionary of the English Language, from http://www.etymonline.com/.

*Geeraerts, D.* (1983). *Reclassifying semantic change.* Oxford: Clarendon.

*Kennedy C., McNally L.* (2005). Scale Structure, Degree Modification, and the Semantics of Gradable Predicates. *Language, 82*(2), 345–381.

*Khalina N.V.* (1993). Категория градуальности в слове и тексте. Барнаул: Алтайский университет.

*Kiryan A.D.* (1981). *Градация как способ организации лексико-семантической группы.* Ph.D. dissertation, Department of Russian, Moscow.

*Kolesnikova S.M.* (2010). Функционально-семантическая категория градуальности в современном русском языке. Москва: Высшая школа.

*Kostusyak N.M.* (2002). Категорія ступенів порівняння прикметників і прислівників. Луцьк: Вежа.

*Kustova G.I.* (2000). Когнитивные модели в семантической деривации и система производных значений. *Вопросы языкознания*, 4, С. 85–109.

*Marchuk L.M.* (2008). *Категорія градації в сучасній українській літературній мові*. Ph.D. dissertation, Department of Ukrainian, Institute of Ukrainian, the National Academy of Sciences of Ukraine, Kiev.

*Osgood Ch.E., Suci G.J., Tannenbaum P.H.* (1987). *The measurement of meaning.* Urbana, Chicago and London: University of Illindis Press.

*Podufalova T.V.* (2007). *Когнітивно-семантичні та функціональні особливості градації в російській мові.* Ph.D. dissertation, Department of Russian, Kharkiv National Pedagogical University, Kharkiv.

*Rasmussen J.E.* (1996). Futher rules of sirenic vowel gradation. *Acta linguistica hafniensia*, 28, 119–124.

*Sapir E.* (1944). Gradience. A study in semantics. *Philosophy of science*, *II*(2), 65.

*Sheygal H.Y.* (1990). Градация в лексической семантике. Куйбышев: Куйбыш. гос. пед. ин-та.

*Shmelev D.N.* (1973/2008). Проблемы семантического анализа лексики (на материале русского языка). Москва: Наука.

*Shvedova N.Yu.* (2005). Русский язык: избранные работы. Москва: Языки славянских культур.

*Stern, G.* (1931). *Meaning and Change of meaning with special reference to the English language.* Göteborg: Elander.

*Stevenson A.* (2007). The Shorter Oxford English Dictionary (6th ed.). New York: Oxford University Press.

*Ullmann, S.* (1962). *Semantics: An introduction to the science of meaning.* Oxford: Blackwell.

Iryna V. Malynovska
*Post-doctoral student, National Academy of Sciences of Ukraine, Ukraine*

# 'Shuttle' methods in the analysis of metaphor in English philosophical discourse

**Abstract:** Metaphor in common language has been given extensive elucidation, while metaphor in philosophical discourse still remains outside much linguistic research and needs special interpretation. Philosophical texts abound with conceptual metaphors, though the amounts differ depending on individual styles. This corpus is mostly represented by non-signaled metaphorically-used words. The data was collected following an identification procedure as defined by the Pragglejaz Group and analyzed with a number of instruments under the umbrella name of 'Shuttle' methods, implying a significant degree of multiple cross-domain mapping. The study contributes to a better understanding of the lingual nature of metaphor as a tool of philosophical reflection.

Keywords: metaphor, types of metaphors, identification of metaphor, function of metaphor in philosophical discourse, cross-mapping.

## 1. Introduction

There is a well-known dictum of F. Nietzsche's "every word is a prejudice" (Nietzsche, 2001). It expresses the pursuit of absolute knowledge alongside a lack of faith in the heuristic abilities of language due to its metaphorical power allegedly preventing statements from being true. This position is still shared by many philosophers and linguists. *The purpose* of my research work is to examine both the abovementioned and contrary approaches, and through the development of position examined in this paper, substantiate a reliable identification procedure for metaphors in philosophical discourse *per se*, as well as address the nature, place and functions of metaphor in English-speaking philosophical discourse. The research corpus was the metaphors offered by textual material. I analyzed the texts written by famous contemporary British and American philosophers belonging to different philosophical traditions and schools (inter alia, Bernstein, 1983; Dennet, 2010; Nagel, 1991; Rorty, 2003; Woodruff, 2001).

## 2. Review of literature

As a philosophical and rhetorical problem, metaphor appeared in Aristotle's speculation which denoted metaphor as displaying creative and artistic power: "…it is metaphor that gives people the pleasure of getting new knowledge" (Aristotle, 1978, p. 115–116). The subsequent development of views on metaphor was marked by the rivalry between this and contrary ideas, which, in its turn, contributed to the evolution of parallel methodological paradigms, from purely speculative to using empirical and experimental methods.

Analytic philosophy formulated the Comparison Theory (Davidson, 1984) which considered metaphor to be "a thing in the head", a mental object preventing clear understanding of utterance and discourse (Peres, 1998). Criticizing such understanding, Marx Black sarcastically noted that metaphors are unwanted by analytic philosophy, albeit both philosophical reflection and writing are inherently metaphorical which expands the ability of the human mind to search and discover (Black, 1962, p. 25).

Following philosophy, linguistics has always had metaphor as a research subject, either as a figure of speech or as a basic cognitive item. The cognitive theory of metaphor does not deny logical attempts of earlier scholars; rather it clarifies details, deepening our understanding of this phenomenon and due to the introduction of semantic modeling, integrates the approaches of rhetoric, semantics, stylistics, sociolinguistics and structural theory. To-day, metaphor is generally interpreted as a universal mentifact which opens a way to the understanding of consciousness and thinking (Kövecses, 2002; Johnson & Larson, 2003; Lakoff & Johnson, 1980, 2003; Wierzbicka, 1971).

Anna Wierzbicka was one of the first to offer a reliable procedure – a two-fold way of identifying metaphors linguistically. According to her method, there are two major characteristics of metaphor. The first is a semantic interpretation, when one of the components of its deep structure should be understood as '… that it is not … but …' (Wierzbicka, 1971). The second is formal interpretation, which implies absolute, or at least partial, ellipsis in the surface structure of this formula.

Similar understanding evolved in a number of linguistic and philosophical schools of the late 20th century. Ivor Richards argued that that language is *vitally* metaphorical (Richards, 1965). In terms of the fundamental theory elaborated by George Lakoff (Lakoff & Johnson, 1980), metaphor is understood as cross-domain mapping between two conceptual areas. The procedures for identifying metaphor offered by these scholars work well with metaphors taken as individual mentifacts in the form of A is a B statement. However, it is too general to achieve a detailed cognitive model of more complicated, semantically overlapping multilevel textual/ discursive metaphors.

An attempt to overcome the complexity of the semantic structure of metaphorical texts was made by Gerard Steen together with other members of the Pragglejaz Group, who developed a five-steps analysis as applied to poetry (Steen, 2009). Their framework evokes a simultaneous movement of two shuttles, one of which scurries back and forth, and the other -in depth and upwards to add contextual textures, interlocking 'visible and invisible' semantic loops of a text, both on the surface and in depth. They continually engage the juxtaposition of horizontal and vertical associations, building a three-dimensional model of the cognitive area of a metaphorical discourse fragment.

## 3. Empirical methods as a part of methodology employed in analysis

The development of theories of metaphor and methods of its study shows that abstract theorizing and logical procedures, taken alone, are not enough to provide the explanation necessary to clarify the nature and role of metaphor, neither in language nor discourse. Dealing with an enormous number of facts, linguistics is generally recognized as a primarily empirical science with a great arsenal of duly approved empirical research methods which is being constantly updated. This arsenal supplies a researcher with a number of options, the combination of which can efficiently fit the objectives of a specific research topic and the specificity of its subject.

Approaching the analysis and methods to identify metaphors in philosophical discourse, I proceeded from a number of methodological assumptions and took corresponding methodological steps:

1. The primary analysis involves *observation of linguistic data as a basic empirical method in linguistics.* Observations recorded natural and not construed linguistic facts. These were metaphors selected from 250 pages of texts for each of the abovementioned authors totaling some 6000 paper or electronic pages. The metaphors found belong to different categories: cognitive, conceptual, root, nonlinguistic and mixed. The average proportion of their usage is 68 to each 500 words (both notional and syntactic) or 13,6%, though the figure varies depending on individual styles. These differences become minimized due to the extensive use of root and dead metaphors.
2. A valid study of a linguistic phenomenon should be based on *observation of speech activity (performance)* manifested in the form of text. Hence, linguistic

facts in this research are understood as representations of metaphorical concepts both in language and texts.

3. Reliable linguistic data can only be obtained with a large array of available texts. That is why in order *to verify* my local observations, in addition to analyzing texts of the chosen philosophers, I referred to the instruments offered by Corpus Linguistics – the academic sections of the Corpus of Contemporary American English (COCA) and the British National Corpus (BYU-BNC). Work with these Corpora was based on Metaphorical Pattern Analysis (Stefanowitsch, 2006).
4. In this study, *classification* as an empirical method is represented indirectly as it was borrowed from a well-known previous research (Steen, 2009).
5. Leaving aside the ongoing discussion on *experiment* in linguistics, I share an understanding of the so-called "*side experiment*" ( also referred to as "*secondary experiment*") as a specialized method of empirical research which includes modeling. In this study, modeling of cognitive structure of metaphor in philosophical discourse was performed.
6. *Measurements* as important empirical tools were applied to define the parameters of metaphor usage in a philosophical text. At this stage of analysis, they were performed in an arithmetic form without applying mathematical statistics in view of the limited number of texts and lexical units.

The abovementioned collection of empirical methods makes for a substantial part of a methodological system as employed in this study, in which theoretical and empirical procedures follow and complement each other at the relevant stages of research (Talmy, 2003, p. 5). In brief, the general line of my strategy can be explicated by:

- formulation of a hypothesis of conceptual metaphor in philosophical discourse introspectively
- observation of texts and sampling probable metaphorical linguistic expressions representing philosophical concepts (in this article the concept of TRUTH is taken for illustration)
- classification of them into target domain and source domain expressions
- verification of these findings by comparing them with the Corpora data
- identification of conceptual philosophical metaphor via modeling its cognitive structure (detailed description is given in the next section).

## 4. Findings and analysis

Addressing any professional discourse requires understanding at least the basic characteristics of its subject area. Philosophy is a very specialized field, substantially differentiated from other areas of human knowledge. It systematically studies most general, fundamental and ultimate problems of human life and operates with universal concepts (megaconcepts) such as HUMAN PURPOSE, EXISTENCE and REALITY, TRUTH, MIND, KNOWLEDGE and REASON, LANGUAGE, VALUES, SOCIETY and MORALITY (Grayling, 1999; Teichmann & Evans, 1999) which are represented by the linguistic units of General English. The sharing of its vocabulary with common language leads to a particular complexity and ambiguity in the language of philosophy, which was explicitly characterized by Ludwig Wittgenstein: "Philosophy can only be written as a literary composition" (Wittgenstein, 1980, p. 24). Metaphors have always been a an integral part of most European philosophical texts, even those committed to a scientistic approach. The functionality of metaphor in such texts is twofold: it performs a degree of aesthetic functions in a manner akin to every other 'poetic composition' while exercising a very important explanatory function. Steven Pepper argued that relatively superficial uses of metaphor in philosophy occur when figures of speech are used instead of exact definitions of newly-born ideas (Pepper, 1982).

There are also permeating uses of metaphors which takes place when such figures of speech are the accepted names of philosophical theories, methods and models: *armchair theorizing, armchair philosophizing, or armchair scholarship* (S.F. Nadel's terms); *reductive realism, naive realism, greedy reductionism* (D. Dennet's terms), the coining of William James' *stream of consciousness*, and Pepper's *root metaphor theory*. Obviously, these nominative units testify to the great explanatory power of metaphor. They are coined when neither common nor professional languages have the resources to fit new ideas or the personal views of their creators.

While analyzing such units, we have to involve a third 'shuttle' in order to cover the theoretical context and socio-cultural background, as well as individual vision implied in a metaphorical text.

Let's consider the following extract:

> "'(Pa) Once the term 'consciousness' is seen to allude to an incompatible congeries of features, and these features are sorted out and described, (Pa') many of the most stubborn perplexities in philosophy of mind dissolve. (Pb) The quest for a plausible and consistent analysis of consciousness develops into the hunting down of that elusive quarry, the little man in the brain, (Pb') who is driven first from his role as introspector only to reappear as perceiver, reasoner, intender and knower… (Pc) Expelling him from our thinking about mind requires, I hope to show, more radical alterations in our views of mental

phenomena than are usually envisaged. (Pd) It is one thing to exorcize the ghost in the machine, (Pd') but he can reappear in more concrete form, as, for example, a stimulus-checking mechanism or – as we have seen – as a brain-writing reader, (Pd") and in these guises he is equally subversive" (Dennet, 2010, pp. 11–112).

Deriving an integral underlying conceptual structure from the linguistic form of the metaphoric text, we proceed from its formalized presentation sentence by sentence or, in the case of scattered metaphor, proposition by proposition. Performing this procedure, I followed the abovementioned Dr. Steen's five-steps framework (Steen, 2009) with the necessary modifications relevant to the specificity of philosophical discourse in terms of considering underlying theoretical ideas and socio-cultural context, and thus adding one further step. The results of this modeling as related to the propositions Pa and Pa' are shown in Table 1.

*Table 1: Modeling of cognitive structure of metaphor in philosophical discourse*

| Steps / Text | (Pa)Once the term 'consciousness' is seen to allude to an incompatible congeries of features, and these features are sorted out and described, (Pa') many of the most stubborn perplexities in philosophy of mind dissolve. |
|---|---|
| 1. Identification of metaphor related words | TO ALLUDE<br>DISSOLVE |
| 2. Identification of propositions | Pa: (<u>ALLUDE</u>, TERM)<br>Pa': (<u>DISSOLVE</u>, PERPLEXITIES) |
| 3. Identification of open comparison | Pa: SIM{зFз a<br>[F(TERM) ], [ALLUDE (a)s]}<br>Pa': SIM {зFз a<br>{[F(PERPLEXITIES)], [DISSOLVE(a)]} |
| 4. Identification of analogical structure | Pa: SIM<br>{ [INDICATE/BE AN INDICATION OF (TERM)], [ALLUDE (HUMAN)s]}<br>Pa': SIM {[VANISH (PERPLEXITIES)], [DISSOLVE (CHEMICHAL)s]} |
| 5. Identification of root correlations | Pa: MINDthr \|\|ALLUDE > INDICATE/BE INDICATION OF HUMANs > TERMs\|\|<br>Pa ,: MINDthr \|\|DISSOLVE > VANISH CHEMICALs solids> PERPLEXITYs stubborn\|\| |

| Steps / Text | (Pa)Once the term 'consciousness' is seen to allude to an incompatible congeries of features, and these features are sorted out and described, (Pa') many of the most stubborn perplexities in philosophy of mind dissolve. |
|---|---|
| 6. Identification of cross-domain mapping | Pa: ALLUDE > INDICATE/BE INDICATION OF HUMANs > TERMs<br>*Inferences:*<br>Goal of ALLUDE/MAKING ALLUSIONs > ESTABLISH CORRESPONDENCESbtw/mind/thr > UNDESTAND BETTER mind<br>Pa ':<br>DISSOLVE > VANISH CHEMICALs solids > PERPLEXITIES stubborn<br>*Inferences:*<br>Goal of DISSOLVE > VANISH > DESTROY > CLEAR THE WAY TO understanding |

*Note:*
P stands for proposition
Pa, Pb – individual proposition in sequence
Pa' – proposition' imbedded in Pa
F – is an arbitrary chosen symbol to denote a function in a conceptual target domain
(*a*) in the conceptual source domain
SIM –similarity/correspondence
з – a predicate symbol
Curly brackets are used for showing the necessary individual members of P
Square brackets are used for detailing the contents of P
|| – two vertical lines delineate theoretical areas in question (thr)

The above table needs some explanation and interpretation of the procedural outcomes which may be extrapolated to a whole set of propositions making up a metaphorical structure embedded in the text base (in other terms, conceptual structure) of the analyzed discourse fragment.

Step 1. (Identification of metaphor-related words). In propositions (Pa), (Pa'), according to my comprehension and dictionaries, there are 2 metaphors, the foci of which being 2 metaphor-related words – *to allude and dissolve*, a vehicle for the first being *to allude*, for the second – *dissolve*: humans, not terms allude; chemicals, not perplexities dissolve. There is open comparison between *terms* and *humans*, and *(stubborn) perplexities* and *(solid) chemicals.*

Step 2. (Identification of propositions). Here the linguistic analysis of the source domain and target domain is followed by the transformation of the linguistic expressions into conceptual structures in the form of a series of propositions (Ibid).

Step 3. (Identification of open comparison). The procedure involves focusing on correspondences/similarities between two conceptual domains in each of the analyzed propositions. It is symbolically written that some activity of the function (F) in the conceptual target domain and some entity (*a*) in the conceptual source domain are in the relations of similarity concerning the activity of *the term 'consciousness'* and the alluding activity of some entity *(a)* in the second part of the proposition (Pa) and the activity of (stubborn) perplexities on one hand, and some entity *(a)* in the proposition (Pa') on the other. It is here that the operator SIM (similarity) is explicitly introduced.

Step 4. (Identification of analogical structure). According to G. Steen, this step is most important: it adds to our understanding of mapping as an action to establish analogy between two domains, with simultaneous on-going 'decoding' of their cognitive meaning, 'undressing' it, stripping of specific semantics layer by layer, and adding, if necessary, further abstract conceptual substance. Following the dialectical process of any analysis, this step represents the descent from the concrete to the abstract.

Step 5. (Identification of root correlations). At this stage I undertake identification of the underlying theoretical ideas and/or academic/cultural context which is of importance for philosophy. To understand what is actually (semantically) implied by '*stubborn perplexities can dissolve*', as well as other such metaphors, we should refer to the vertical context of the text. D. Dennet's book "Content and Consciousness" was written in 1969, at the time of the long-lasting phenomenological debate. In 1967 Arthur Koestler, a philosophical psychologist published his treatise under the name of "The Ghost in the Machine", which in its turn had its title borrowed from the Oxford philosopher Gilbert Ryle, who applied it to describe the mind-body relationship. The label 'stubborn perplexities' was addressed to the opponents of the theory stipulating that the human mind is based on primitive atavistic brain structures responsible for aggressiveness, anger, hatred, love and happiness. The metaphors, used in the propositions *(Pb)-(Pd") – hunting down of that elusive quarry, the little man in the brain, exorcize the ghost in the machine, brain-writing reader* – serve both as allusions to the academic context, and units of the professional phenomenological language in the absence of a more systematic terminology. In our modeling these theoretical implications are shown with symbols to delineate the extent to which these contextual uses are valid.

Step 6. (Identification of cross-domain mapping). This step aggregates the earlier pre-models in a more or less complete model of cross-domain mapping underlying the relevant lingual explication. Implicit elements of the human activity of making allusions is projected onto implicit elements of the term scheme and the scheme of *dissolve* is projected on the scheme of *perplexity*.

It goes without saying that the whole meaning of the text is much more complicated and it is only endless multiplication and shuttling reduplication of the procedural steps that can qualify for a comprehensive model of the cognitive processes underlying the composition of this or any other philosophical text.

## 5. Conclusion

Linguists generally agree on the fact that most metaphors are represented in discourse by non-signaled words. This is true for analyzed texts from philosophy. Having in mind the difficulties of metaphor identification, while elaborating an adequate research process I proceeded from two basic interrelated principles: the necessity to explicate both the surface and the underlying structures of metaphor as well as modeling its conceptual structure in a text. The theoretical context based on identification of metaphors showed that cognitive metaphors prevail in the case when they are used for the explanation of some new or debatable theoretical ideas; others are either deeply embedded in philosophical reasoning and form a part of professional language, or may be used for the sake of the aesthetic value of discourse.

The analysis demonstrated that in their writing strategies, modern philosophers share Aristotle's opinion quoted above, and use metaphor as a tool of philosophical reflection.

## References

*Aristotle.* (1978). *The Art of Rhetoric.* Moscow: Antic rhetorics. [In Russian].

*Davidson, D.* (1984). Thought and Talk. In *Inquiries into Truth and Interpretation.* Oxford: Oxford University Press, pp. 70–155.

*Grayling, A.C.* (1999). Editor's Introduction. In *Philosophy 1: A Guide through the Subject* (p. 1). Oxford University Press.

*Johnson, M. & Larson, S.* (2003). Something in the way she moves. Metaphors of musical motion. *Metaphor and Symbol*, 18, 63–84.

*Kövecses, Z.* (2002). *Metaphor: a practical introduction.* Oxford and New York: Oxford University Press, US.

*Lakoff, G.* (1991). Metaphor and War. The Metaphor System Used to Justify War in the Gulf. *Viet Nam Generation Journal,* 3(3). Retrieved November, 1991, from http://www2.iath.virginia.edu/sixties/HTML_docs/Texts/Scholarly/Lakoff_Gulf_Metaphor_1.html.

*Lakoff, G.* (2001). *Metaphorical Thought in Foreign Policy. Why Strategic Framing Matters*. The Frameworks Institute.

*Lakoff, G. & M. Johnson.* (1980/ 2003). *Metaphors We Live By*. Chicago: University of Chicago Press.

*Neitzsche, F.* (2001). The wanderer and his shadow. In *Human, All-Too-Human II* (pp. 172–341). The Project Gutenberg EBook.

*Pepper, S.* (1982). Metaphor in Philosophy. *The Journal of Mind and Behavior,* 3(3–4), 197–205.

*Peres, C.* (1998). On Using Metaphors in Philosophy. In: *Proceedings of the Twentieth World Congress of Philosophy held in Boston, Massachusetts U.S.A., Aug. 10–16, 1998*. Retrieved January 29, 2000 from http://www.bu.edu/wcp/Papers/Meth/MethPere.htm.

*Richards, I. A.* (1965). *The philosophy of Rhetoric*. New York: Oxford UP.

*Steen, G.J.* (2009). From linguistic form to conceptual structure in five steps: Analyzing metaphor in poetry. *Cognitive poetics*. Berlin and New York: Mouton de Gruyter. Retrieved from http://academia.edu/250001/From_linguistic_form_to_conceptual_structure_in_five_steps_analyzing_metaphor_in_poetry.

*Stefanowitsch, A.* (2006). Words and their metaphors: A corpus-based approach. In A. Stefanowitsch & S.Th. Gries (Eds.), Corpus-based approaches to metaphor and metonymy (pp. 63–103). Berlin and New York: Mouton de Gruyter. Retrieved from http://www.twirpx.com/file/992820/.

*Talmy, L.* (2003). Concept structuring systems / Talmy L. Toward a cognitive semantics: in 2 v. – V.1. – Cambridge (Mass.); London: A Bradford Book: The MIT Press.

*Teichmann, J. & Evans, K. C.* (1999). *Philosophy: A Beginner's Guide*. Blackwell Publishing.

*Wierzbicka, A.* (1971). Comparison — gradation – metaphor. *Pamiętnik Literacki*, 4, 127–148. [In Polish].

*Wittgenstein, L.* (1980). *Culture and Value*. Oxford: Basil Blackwell.

## Illustrative references

*Bernstein, R.J.* (1983). *Beyond objectivism and relativism: science, hermeneutics, and praxis*. Pennsylvania: University of Pennsylvania Press.

*Dennett, D.* (2010). *Content and Consciousness*. New-York: Routledge.

*Nagel, T.* (1991). *Equality and partiality.* New-York: Oxford University Press, Inc.

*Talfourd, Thomas Noon.* (1838). *The works of Chareles Lamb to which are prefixed, his letters, and a sketch of his life.* New-York: Harper & brothers.

*Woodruff, P.* (2001). *Reverence: Renewing a forgotten virtue.* New-York: Oxford University Press.

# Part Four: Miscellaneous Methods

Mirosław Bańko
*University of Warsaw, Poland*

# Iconic effects in loanword adaptation*

**Abstract:** The way a loanword accommodates to the recipient language is often guided by a tendency to maintain harmony between its form and its meaning. The tendency can be seen as an example of iconicity in natural languages. Drawing on examples from an ongoing research project, this paper presents a typology of iconic effects that appear in the process of loanword adaptation. A variety of methods used in the project will be shown, ranging from corpus queries to psycholinguistic experiments.

Keywords: iconicity, lexical borrowings, loanword adaptation, synonymy, word perception.

## 1. Introduction

Iconicity in natural languages can be defined as a similarity relation between linguistic signs and their referents. Among signs which are iconic in this sense there are onomatopoeic words, certain prosodic phenomena and gestures, whether in sign languages or in normal, auditory speech. More abstract cases of sign – referent similarity involve resemblance of structure and are known under the term *diagrammaticality*. For example, in compound words the order of elements is indicative of the order of their referents (e.g., a *Polish-English* dictionary can be expected to give Polish words first), the order of verbs in a story corresponds to the order of events (cf. *Veni, vidi, vici*), and the structure of one's CV reflects, albeit in the reverse order, the sequence of events in one's life.

The scope of iconic effects becomes even broader when non-arbitrariness of signs is taken as the defining criterion. For example, the word *wing* does not resemble a part of an aircraft, but it is not wholly arbitrary either: its figurative use is clearly motivated by its primary use and the obvious resemblance between aircraft wings and birds' wings. Likewise, it is no accident that the word *throne* can refer both to a piece of furniture and to royal power. Metaphor and metonymy, illustrated by the examples above, are two important sources of non-arbitrary

* This study was carried out within a Polish-Czech project: *Adaptation, perception and reception of verbal loans in Polish and Czech – linguistic, psychological and historic-cultural factors*, supported with a grant from the National Science Centre, Poland, no DEC-2011/03/B/HS2/02279.

signs and other examples can be cited, e.g., from the domain of folk etymology, blending and analogical formations.

In what follows, iconicity is understood as a property of signs whose meaning stands in a non-arbitrary relation to their form. Non-arbitrary, or iconic, signs may also be considered motivated, because their meaning is motivated by their form. Taking the form as the starting point, one can say as well that the form of iconic signs is expressive, because it suggests their meaning (the notion of expressivity adopted here is different from that of Bühler 2004/1934). Thus, iconicity, non-arbitrariness, motivation and expressivity are closely related concepts which refer basically to the same phenomenon.

Iconic effects can be more or less visible, ranging from relatively simple cases like onomatopoeia to more abstract ones, such as when the connotative meaning of a word is shaped by its pronunciation or spelling. In this paper, I will focus on examples of iconicity which appear in loanword adaptation. These examples come from a research project which has to be shortly described before we can go any further.

## 2. The APPROVAL project

The aim of the project is to identify factors bearing on the adaptation of loanwords. Such factors can be diverse in nature: some of them may be connected with human perception, i.e. relatively culture-independent, others may be related to the structure of the recipient language, while still others may be bound with the writing tradition of a nation or its openness to foreign influences. Seeking to advance beyond general statements like these, we have decided to compare systematically a number of lexical loans with their native synonyms and account for similarities and differences between the foreign and the native element in each pair. Spelling variants of loanwords differing in the degree of adaptation have likewise been taken into consideration. Only synchronic loans have been included, i.e. those whose foreign origin is still visible to average language users. A hundred word pairs (sometimes triples, quadruples, etc.) have been selected for analysis and the results are being published successively on the project website (www.approval.uw.edu.pl).

One half of the material, i.e. fifty word pairs, is drawn from Polish, the other half from Czech. The two halves are parallel in the sense that they are made of translational equivalents, e.g. the Polish pair *absurdalny – niedorzeczny* corresponds to the Czech pair *absurdní – nesmyslný*. In addition, the first element in each Polish pair comes from the same root as the first element in the corresponding Czech pair and often the two words come from the same foreign language. Composing the research material this way made it possible to study the adaptation processes in two languages on the basis of comparable examples, using the same methodology, the same kind of data, and even the same description format. The danger of overgeneralizations was reduced (as compared to research based on one language only), and the opportunity emerged to observe the influence of the normative traditions of Polish and Czech on the way each language adopts foreign vocabulary.

## 3. Review of literature

One research area relevant to this study is concerned with synonymic relations between words. Numerous publications have been devoted to synonymy and, in particular, the question of whether exact (or absolute, or total) synonymy exists, see Bańko and Svobodová (in print) for a review. Much depends on how synonymy is conceived: it is not hard to find two words of the same designative meaning, but even such words are likely to differ in other respects, e.g. their connotative meaning, stylistic distribution, or frequency.

Another context in which the present study can be set is that of iconicity in natural languages. A working hypothesis adopted in the APPROVAL project is that the form of a word is not irrelevant (see also Bolinger 1949, who put it succinctly in the title of his paper: 'The sign is not arbitrary'). This hypothesis is incongruent with the tradition of Saussurian linguistics, although the research aims of the APPROVAL project – in particular, the focus on distinctions between words – are in perfect agreement with the assumptions of linguistic structuralism. After all, as Saussure (1961/1916) taught, language is all based on distinctions.

Some earlier studies suggest that the process of loanword adaptation may be guided, at least partially, by a tendency to maintain harmony between the form of a word and its meaning, see Bańko (2008a, 2008b, 2009, 2013a), Bańko and Hebal-Jezierska (2012). This tendency operates in two ways: first, it can change the meaning of a loanword to adjust it to its form; second, it can change the form of a word to make it more coherent with its meaning. In many cases, however, the dependence between form and meaning is mutual: the form adjusts to the meaning and vice versa, so it is difficult to say where the process begins.

An example of how the meaning of a word can be affected by its form can be found in *kurort*, a 19th-century Polish loan from German, initially used in the original sense 'health resort'. It later changed its meaning to 'popular and snobbish holiday place' not only because a native synonym *uzdrowisko* was coined, but also because the associations the loanword evoked in the minds of the speakers were coherent with its foreign shape and origin. Among them, foreignness has to be mentioned first (the word was and still is most often used with reference to foreign places), but modernity, reputation, elegance and exclusiveness (strangely enough, not in conflict with popularity) are also important (all of these being clearly visible in collocation analysis, see Bańko 2013a).

An example of the opposite process, i.e., the influence of word meaning on word form, can be found in *torreador* 'bull fighter', a peculiarly Polish lexical innovation, in which the letter *r* has been duplicated against etymology to make the word more expressive (Bańko 2008a). Another example is *jazz*, almost a hundred years after being borrowed into Polish and Czech still used much more often in its original spelling than in the Polonized or Bohemized one, largely because the native spellings *dżez* and *džez*, respectively, are less suitable to convey the historic and cultural connotations of the word (Bańko and Hebal-Jezierska 2012).

## 4. Research methods

When looking for differences between exact synonyms or variant forms of a word, it is necessary to adopt a broad notion of meaning, comprising not only the word's designative value, but also its connotative features, typical collocations, stylistic distribution, emotive overtones, etc. Leech's (1981: 9–23) early proposal of dividing meaning into several components was adopted as a starting point in the APPROVAL project, but it was later adjusted to the project's purposes and the data available.

The research has taken on the form of case studies, based on dictionaries and corpora, sometimes also other sources, e.g. Google images and library catalogues. The gallery of images assigned to a given word by the Google search engine tells something about the words' place in mass culture, and often reveals connotations which are otherwise not easy to recognize (see Bańko 2013b). Likewise, the relative frequencies of two words in book titles in an online library catalogue can provide additional data concerning the words' scope of usage, thus complementing the corpus analysis.

In parallel to the linguistic analysis of word pairs, psycholinguistic experiments are carried out to discover the differences between the same words and measure the semantic distance between them. Some of the experiments are based on Osgood's semantic differential, others are to be performed in a semantic space constructed on the basis of the National Corpus of Polish (see Rączaszek-Leonardi 2013 for details). So far a free association study has been carried out and the results are available on the project website. In other studies, which have not been completed yet, the subjects are asked to judge particular words along predefined dimensions, resulting from corpus analysis, or to respond in a certain way to sentences in which such words are embedded. This way we hope to find out whether, for instance, the choice of the English spelling *chipsy* 'crisps' instead the Polonized phonetic spelling *czipsy* has some influence on how Polish speakers evaluate the snack on the tasty vs. tasteless or healthy vs. unhealthy scales.

By combining linguistic and psycholinguistic methods, performing analyses on two languages and using different kinds of data, both textual and iconographic, more valuable results can be obtained than otherwise possible. The project is still in progress: the case studies have not been completed, the psychological experiments are halfway, the findings from the linguistic and the psycholinguistic parts of the project still need to be systematically compared, and a summarizing analysis has yet to be performed. Nonetheless, we can already briefly present some of the results. Two examples from Polish will be given first, followed by a tentative typology of iconic effects. For lack of space, Czech material will not be used here.

## 5. Examples

### 5.1 *strofa – zwrotka*

Both words correspond to the English term *stanza*. *Strofa* – coming from the same Greek root as English *strophe* (though borrowed from Latin) – is older, used as early as in the 17th century. *Zwrotka*, almost an exact translation of *strofa*, was coined presumably in the second half of the 18th century, the period when Polish scientific terminology began to develop. For a long time both words were equally widespread in literary works and in specialist literature, but around halfway through the 20th century *strofa* started to predominate in scholarly texts.

Nowadays *strofa* belongs to a higher register, being a literary word or a technical term, while *zwrotka* is stylistically unmarked and used in general language. For example, *zwrotka*, but not *strofa*, is used in reference to popular songs and children's poems. As collocation analysis indicates, *strofa* is the subject of aesthetic evaluation (cf. *piękne strofy* 'beautiful stanzas') and artistic activity (cf. *pisać, układać strofy* 'write, arrange stanzas'), while *zwrotka* is less frequent in such contexts; in addition, *strofa* can be recited, but *zwrotka* is sung. Only *zwrotka* collocates with the word *refren* ('refrain'), which confirms its connection to songs.

The results of corpus analysis are in line with Osgood's free association test. Among the associations noted more than once, *refren* 'refrain', *rymy* 'rhymes', *śpiewanie* 'singing', *muzyka* 'music' and *ognisko* 'camp-fire' were only given for *zwrotka*, while *szkoła* 'school', *poezja* 'poetry', *literatura* 'literature' and *Mickiewicz* (the best known Polish poet) were only given for *strofa*. In addition, though *piosenka* 'song' and *wiersz* 'poem' were mentioned for both words, *piosenka* had the frequency of 19 with *zwrotka* and 3 with *strofa*, while *wiersz* appeared 16 times with *strofa* and only 3 times with *zwrotka*. (The maximal number of responses in this test was 66, with up to 3 associations from each of 22 subjects.)

The distinctions in use observed between *strofa* and *zwrotka* correspond to the distinctions in form, and are likely to result from them. The strange shape of *strofa*, non-transparent to laymen, contrasts to the familiar shape of *zwrotka*, easily recognized as related to other Polish words, e.g. *zwrot* ('turn'), and having the appearance of a diminutive (cf. its pseudo-diminutive suffix -*k*-). Moreover, the strange shape of *strofa* creates a distance and is a good vehicle for the content suggested by the word: erudition and technical knowledge. The familiar form of *zwrotka*, on the other hand, evokes a feeling of conceptual and emotional closeness of the referent.

## 5.2 *dealer – diler*

This word – having two spelling variants in Polish – was borrowed from English in the late 1980s or early 1990s. Like many other lexical loans, it has a narrower scope of usage than in the donor language: it is applied mainly to car vendors, especially in authorized showrooms, as well as to drug sellers.

Its variants have different stylistic distribution and, even more interestingly, partly different reference. They are both most often used in the press, but *dealer* is almost twice as frequent there as *diler* and refers mainly to car vendors, whereas *diler* has equal frequency in automobile and drug contexts. In literary texts, on the other hand, *diler* is twice as frequent as *dealer* and is applied usually to drug sellers, whereas *dealer* is equally often used in automobile contexts. In addition, though the overall frequency of *dealer* is higher, *diler* predominates when it comes to drugs.

The predominance of *dealer* in journalism and *diler* in fiction is no surprise: as a rule, the authors of literary texts want to involve their readers emotionally, so they avoid foreign words and they avoid foreign spellings too, if there is a choice between an assimilated and an unassimilated variant. Foreign words and foreign variants would create a distance between the readers and the plot, an effect most writers would like to avoid.

It is much more interesting that *dealer*, especially in the press and in technical literature, more often refers to car vendors than drug sellers. A likely explanation is that the foreign variant is more prestigious and better suited to name the job of authorized vendors in car showrooms than the illegal and socially stigmatized activity of people who sell drugs. The domestic variant, on the other hand, is unpretentious and corresponds well with the suspicious job of drug sellers.

A glance at Google images corroborates the observations made in corpus analysis: the first hundred images for *dealer* show a mixture of casinos, cars and drugs, whereas the images for *diler* show drugs almost exclusively. The same difference can be seen, although less clearly so, in Osgood's free association test, e.g., *samochody* 'cars' were associated only with *dealer* and *więzienie* 'prison' only with *diler* (but *narkotyki* 'drugs' were given frequently for both words).

## 6. Iconic affects in loanword adaptation: a typology

As can be seen from the examples above, some details of loanword adaptation are best explained by assuming that some kind of iconicity is involved in the adaptation processes. Let us now distinguish several types of iconic effects in loanword adaptation in order to show different ways in which the meaning of a lexical loan can be motivated by its form. The list is by no means exhaustive, it simply comprises what has been observed so far. As noted earlier, meaning is conceived in a broad sense here, including both designative and non-designative features.

### 6.1 Foreign names – foreign realities

Many loanwords refer to foreign realities, e.g. places, people, objects and customs. Such words, known as exotisms, exemplify the meaning – form harmony in a very straightforward way: a foreign name has a foreign referent. Significantly, exotisms are not normally substituted with native words: even the most adamant purists do not find it necessary, probably because the use of foreign names with reference to foreign realities is desirable for both communicative and stylistic reasons. For example, in reportages and travel literature foreign names add to the flavor of the places described.

Even when a loanword has a native synonym in the recipient language, they may differ in terms of where their referents are. The *kurort* – *uzdrowisko* synonym pair can be mentioned here again, because *kurort*, a 19th-century Polish loan from German, was initially most often used with reference to health resorts in Germany (see *Słownik wyrazów obcych* from Michal Arct's publishing house of 1921). If there had been no similar places in Poland, purists would not have insisted on using the native coinage *uzdrowisko* instead. Eventually, the two words diverged in meaning, but even today *kurort* is a few times more frequent in reference to holiday resorts abroad than to those in Poland, while *uzdrowisko* is most often used with reference to Polish health resorts.

### 6.2 Foreign names – technical knowledge

Another way in which the meaning of a loanword can be made congruent with its form is exemplified by the synonym pair *strofa* – *zwrotka*. As we saw above, the two words were first identical in meaning and use, but around half a century ago began they to differentiate: now *strofa*, being recognized as a loanword, tends to be used in specialist language, while *zwrotka*, its native synonym, prevails in general speech. The same difference can be observed between *toksyczny* 'toxic' and *trujący* 'poisonous' (see the APPROVAL project website), as well as in many other synonym pairs. Generally speaking, synchronic loans are perceived as formal, bookish or even erudite, which makes them appropriate in specialized languages, whether in science or the humanities.

There are exceptions to this rule, arising from purist reforms of a language or a language variety. For instance, Polish medical terminology became Polonized to such an extent in the 19th and early 20th centuries that in the pair *puls* – *tętno*, both meaning 'pulse', the native element is more often used in specialist texts and the foreign one more frequent in the general language, against the general tendency.

### 6.3 Foreign names – more prestige

The next type of form – meaning relationship can be illustrated by means of the *dealer* – *diler* variant pair (see above). The word comes from English, a prestigious language in Poland, so it is perceived differently in its original spelling than when spelled phonetically. The difference is not always as clear as here – one variant tending to refer to car vendors, the other to drug sellers – but as a rule, loans from prestigious languages are more likely to promise something exclusive, elegant, extraordinary or simply of higher quality when they retain their foreign spellings. The same holds for synonym pairs: a loan from a prestigious language seems to add more prestige to its referents than a word with exactly the same designative meaning but of domestic origin. Some linguists would complain here about the snobbery of language users or the negative influence of linguistic fashion, but I will not pick up this thread now, because it deserves more attention than I am able to pay to it in this paper.

### 6.4 Foreign names – more emphasis

The Polish terms *eksplozja* 'explosion' and *wybuch* 'explosion, outbreak, outburst' may refer to the same kind of events, but the phenomena referred to by the former word are perceived as stronger and more violent (see the project website).

There are more synonym pairs in which the referents of a loanword seem larger and more powerful than the referents of its native synonym, cf. another pair from the project website: *dewastować* – *niszczyć* (both meaning 'destroy'), in which the first element, cognate with English *devastate*, denotes a purposeful or mindless activity, especially against public property or natural environment. However, sometimes the difference is not in referents themselves, but in the way they are talked about, e.g. in the values conveyed. For instance, the adjective *kuriozalny* 'peculiar, bizarre', related to Latin *curiosum*, is half as frequent in Polish as its native counterpart *osobliwy* 'peculiar', but in parliamentary reports the former word prevails overwhelmingly. A closer inspection shows that Polish MPs use it to criticize their political opponents, e.g. *Pana poglądy są kuriozalne, panie pośle* 'Your views are bizarre, Mr. X'.

As before, there are exceptions to the 'foreign names – more emphasis' pattern. One well-known example is the neo-Latin term *Euthanasie*, used euphemistically in Nazi Germany for the killing of mental patients, i.e. to conceal the truth. Another example can be found in words connected with defecation and urination, often replaced with technical or old-fashioned terms, cf. Polish *urynał* (of French origin) or English *chamber pot* (partly from French), both referring to the same object.

## 7. Conclusions

The tendency to maintain harmony between form and meaning in the process of loanword adaptation probably has some psychological basis. As synchronic loans are usually less familiar than their native synonyms, they probably require more effort to be processed in the mind. For the same reason, non-assimilated variants are likely to require more effort than their assimilated equivalents. So is it this processing fluency that underlies the different perception of particular words and even affects their meaning and use?

There is no definite answer to this question yet, but experiments described by Song and Schwarz (2010) do throw some light on it. These authors experimented with nonce words, some of them familiar in shape, some strange, and observed a correlation between the familiarity of a word and its perception. For instance, fictitious food additives with names difficult to pronounce were evaluated as more harmful than food additives with easy names. Song and Schwarz explain the effect with a mistaken projection of the difficulties the subjects experienced in processing the words onto the referents of the words: unaware of the source of difficulty, the subjects attributed it to the referents, judging them as more risky, more dangerous, more harmful, etc. (for a critical review of these studies see Rączaszek-Leonardi 2013).

Finding an explanation for different examples of iconicity in natural languages has always been troublesome. All the more important it, therefore, is to seek a psychological basis for the kind of iconic effects presented in this paper.

## References

APPROVAL: Adaptation, perception and reception of verbal loans in Polish and Czech – linguistic, psychological and historic-cultural factors. Retrieved September 1, 2014 from http://www.approval.uw.edu.pl/en_GB/.

*Bańko, M.* (2008a). Dlaczego *torreador*? [Why *torreador*?] In M. Kita (Ed.), *Błąd językowy w perspektywie komunikacyjnej*. Katowice: Wyższa Szkoła Zarządzania Marketingowego i Języków Obcych, 297–309. Retrieved September 1, 2014 from http://www2.polon.uw.edu.pl/banko/inne_publikacje.html.

*Bańko, M.* (2008b). Dlaczego *wow*? [Why *wow*?] In G. Dąbkowski (Ed.), *Reverendissimae Halinae Satkiewicz cum magna aestimatione*. Warszawa: Plejada, 9–21. Retrieved September 1, 2014 from http://www2.polon.uw.edu.pl/banko/inne_publikacje.html.

*Bańko, M.* (2009). Should the computer mouse click differently in Polish? In Ch. Voss & A. Nagórko (Eds.), *Die Europäizität der Slawia oder die Slawizität Europas. Ein Beitrag der kultur- und sprachrelativistischen Linguistik*. München – Berlin: Verlag Otto Sagner, 177–185.

*Bańko, M.* (2013a). Normatywista na rozdrożu. Dwugłos w sprawie tzw. kryterium narodowego. [The prescriptivist at the crossroads: discussion on the so-called national criterion] In J. Migdał & A. Piotrowska-Wojaczyk (Eds.), *Cum reverentia, gratia, amicitia... Księga jubileuszowa dedykowana Profesorowi Bogdanowi Walczakowi*, vol. 1. Poznań: Wydawnictwo Rys, 141–148.

*Bańko, M.* (2013b). Obrazy Google jako źródło informacji lingwistycznej. [Google images as a source of linguistic information] In W. Chlebda (Ed.), *Na tropach korpusów. W poszukiwaniu optymalnych zbiorów tekstów*. Opole: Wydawnictwo Uniwersytetu Opolskiego, 9–17.

*Bańko, M., & Hebal-Jezierska, J.* (2012). Proč *jazz*, nikoliv *džez*? Harmonie grafické podoby lexému a obsahu – jako jeden z činitelů ovlivňujících adaptaci cizojazyčných přejímek? [Why *jazz*, and not *džez*? Form – meaning harmony as one of the factors affecting the adaptation of loanwords] In S. Čmejrková, J. Hoffmannová & J. Klímová (Eds.), *Čeština v pohledu synchronním a diachronním*. Praha: Karolinum, 371–375.

Bańko, M, & Svobodová. D. (in print). The role of form-meaning relationship in the process of loanword adaptation. *Polonica*.

*Bolinger, D. L.* (1949). The sign is not arbitrary. *Boletin del Instituto Caro y Cuervo*, 5, 52–62. [Reprinted in: Bolinger, D. L., *Forms of English. Accent, Morpheme, Word Order*. Cambridge, Mass.: Harvard University Press, 1965, 231–239].

*Bühler, K.* (2004). *Teoria języka. O językowej funkcji przedstawiania*. Kraków: UNIVERSITAS. [Theory of language: the representational function of language] (Original work published in German 1934).

*Leech, G.* (1981). *Semantics*. Harmondsworth: Penguin Books.

*Rączaszek-Leonardi, J.* (2013). Studying the semantics of loanwords which have near synonyms in the host language: Psycholinguistic and multidimensional corpus representation methods. Retrieved September 1, 2014 from http://www.approval.uw.edu.pl/en_GB/publikacje.

*Saussure, F. de* (1961). *Kurs językoznawstwa ogólnego*. Warszawa: PWN. [Course in general linguistics] (Original work published in French 1916).

*Słownik wyrazów obcych. 25.000 wyrazów, wyrażeń, zwrotów i przysłów cudzoziemskich, używanych w mowie potocznej i w prasie polskiej* [Dictionary of loanwords: 25,000 words, expressions, phrases and foreign proverbs used in everyday speech and in press in Poland] (9th ed.). (1921). Warszawa: Wydawnictwo M. Arcta. Retrieved September 1, 2014 from http://archive.org/details/sownikwyrazwob00arctuoft.

Katarzyna Stadnik
*Maria Curie-Skłodowska University, Lublin*

# Chaucer's selected narratives: through the looking-glass of medieval imagery

**Abstract:** The paper addresses the issue of the situatedness of the language user vis-à-vis the conception of language as a channel of cultural transmission. This research perspective takes as its point of departure the problem of the continuity of imagery in language and culture, where culture is rendered in terms of the community's conceptual order that allows its members to make sense of experience. While situated nature of human cognition is posited, attention is paid to the role of external representations encapsulating imagery shared in the community, as exemplified by historical data from Chaucer's *Knight's Tale* and *Troilus and Criseyde*, as well as (late) medieval visual culture. Overall, it is argued that both language and artefacts from visual culture may function as repositories, or memory carriers of the community's symbolic tradition, enabling transmission of conceptual knowledge.

Keywords: language and culture, external representations, imagery, memory, situated cognition.

## 1. Preliminary remarks

The paper addresses the issue of language-culture relation from a broadly defined Cognitive Linguistic (CL) perspective. Lately, the significance of research into the situated nature of cognition has been recognised by CL scholars working within the framework of sociocultural situatedness (Frank, 2008). Assuming the situated character of human cognition, we examine the question of how language and non-linguistic cultural resources may facilitate transmission of knowledge. The general problem is exemplified with Chaucer's *Knight's Tale* (KT) and *Troilus and Criseyde* (TR), inspired by Boethius' *Consolation of Philosophy* (Mann, 2003).

In particular, we focus on the issue of how abstract concepts might be transmitted across generations of language users within a particular cultural community. The key problem pertains to the representation of such concepts. Given their lack of perceptual features, the question of grounding abstract concepts in sensory-motor experiences arises. Although in CL the currently dominant view on the problem seems to be that of conceptual metaphor, some empirical findings suggest that abstract concepts may well be represented by concrete situations (Pecher, 2013). Since conceptual knowledge may have perceptual and linguistic origins (Boroditsky & Prinz, 2008), our paper addresses the role of visual culture relative to the involvement of language as one source of conceptual knowledge.

In what follows, we review relevant literature, including crucial aspects of Chaucer's sociocultural situatedness and related empirical research. While discussing selected imagery in KT and TR, we explain why we view external representations from medieval visual culture as complementary to language in the transmission of conceptual knowledge across generations of a cultural community.

## 2. Language and external representations: image as a memory carrier

As indicated, our research falls within the framework of situated cognition. The idea can be theorised differently, depending on the interpretation of what constitutes situational context (Robbins & Aydede, 2009). The embodiment thesis posits that cognition entails not merely the brain, but also the body. Taking the issue a step further, cognition is viewed as embedded insofar as "cognitive ability routinely exploits structure in the natural and social environment" (Robbins & Aydede, 2009, p. 3). The extension thesis posits that "the boundaries of cognitive systems lie outside the envelope of individual organisms, encompassing features of the physical and social environment" (Robbins & Aydede, 2009, pp. 7–8). Since we explore the issue of continuity of culture as the community's conceptual order, we present a view of cognition that transcends the tripartite division, investigating how language and visual culture may complement each other and foster cultural transmission.

Thus, in accordance with the extended cognition perspective, resources of visual culture are seen as external representations. As Kirsh (2010) argues, an "externalized structure can be shared as an object of thought" (p. 444). The physical persistence over time also affects the "shareability" of such external vehicles for thought (Kirsh, 2010, p. 448). As such, external representations may constitute memory carriers comprising information about the past (Golka, 2009), and might enable transmission of knowledge formative for the community's future (Hałas, 2012, p. 170). Therefore, they seem important for the maintenance of the community's conceptual order.

Fundamental to our analysis is the assumption that human conceptual knowledge can be stored in various resources, not only language. Specifically, it may derive from artefacts of visual culture. If so, what is the role of imagery in medieval culture? The answer may inhere in Chaucer's sociocultural situatedness, which "incorporates the need to consider the specific, concrete sociocultural situation in which the individual's cognitive activity is to take place" (Bernárdez, 2008, p. 150).

Indeed, "cognition is not implemented solely within the brain but ... extended by aspects of a person's physical and social context, including other people" (Smith & Collins, 2010, p. 126). Thus, human cognition can be seen as distributed,

> not contained within minds, but implemented by systems that link minds with aspects of the physical and social environment.... much of our distributed cognition is actually distributed across other people. Such a distribution occurs whenever people establish and maintain a socially shared system of meaning. (Smith & Conrey, 2009, p. 461)

It follows that the role of images in transmitting conceptual order of a cultural community should be considered relative to the medieval view of the human mind. Indeed, imagination was considered as an important aspect of human thinking, and cognition was believed to entail the ability to "make a mirror of one's mind" (Carruthers, 2013, p. 74). The human mind was one possible repository of memory, hence the idea of "the learned person as a living library, one who makes a mental chest of memorized texts and materials" (Carruthers, 2008b, pp. 1–2). In this light, Richard de Fournivale's description of the House of Memory succinctly explains the importance of image as a means of access to knowledge gathered by previous generations,

> to this same end, one can come either by painting or by speech. For when one sees a story painted, whether a story of Troy or of some other thing, one sees the deeds of the brave men who were there in past times as if they were present.... when one hears a tale read, one perceives the wondrous deeds as if one were to see them taking place. And since what is past is made present by those two means ... it is clear that by these two things one can come to remembrance. (in Kolve, 1984, p. 25)

It was believed that "conceptions (*imagines*) are ... constructed by the mind from all the materials of sensation" (Carruthers, 2008b, p. 5). Thus, imagery could persist thanks to the cultivation of the craft of memory not only via linguistic channels, but also external resources of contemporary visual culture. Serving as a "memorial hook and cue" (Carruthers, 2008a, p. 274), image was also a means of knowing (Kolve, 1984), a view of the human mind to which Chaucer's *Boece*, the poet's translation of Boethius' *Consolation of Philosophy*, seems to testify (Brown, 2011, pp. 177–182). If so, language and visual culture effectively complemented each other as memory carriers of imagery. Due to their continuity in time, language and visual culture enabled the gradual accumulation of conceptual knowledge in various resources, which meant the accrual of layers in the imagery of the community's symbolic tradition (see also Zalewska, 2014). Ripa's *Iconologia* seems a relevant example illustrating the palimpsest nature of the symbolic tradition. Overall, this development of imagery might be rendered in terms of cumulative cultural evolution (Tomasello, 2014). It is characterised by the "ratchet effect" in that communities do not build their inventories of shared imagery from scratch, but inherit them from previous generations.

## 3. Empirical methodology: establishing situational patterns behind imagery

As already indicated, our research falls within the framework of Cognitive Linguistics. Thus, while we consider language as derivative of human cognition, we also subscribe to the CL methodology of converging evidence, which means that linguistic theorising must be "consistent with what cognitive scientists know about other areas of cognition" (Evans and Green, 2006, p. 17). Accordingly, since cognition may be seen as situated, we seek to establish how the sociocultural situatedness of the language user may find its way into usage. Insofar as language is one type of repository of cumulative knowledge, its evolution may be correlated with cumulative cultural evolution. Our CL-oriented investigation of the historical data selected for illustration of the problem is expected to reveal how evolution of a cultural community's conceptual order may leave its traces in language.

While image served as a memory carrier encapsulating community's cumulative knowledge, what calls for increased attention is the "the medium-specific nature" of encoding information (Kirsh, 2010, p. 446). As a "visual poet" (Brown, 2011, p. 183), Chaucer could evoke relevant images from the symbolic tradition in the minds of his audiences, knowing that both the listeners and readers were familiar with them via literature and visual culture. Therefore, we address the question of how knowledge may be represented so as to effectively sustain the community's conceptual order, and allow its evolution.

In accordance with the empirical methodology of converging evidence, we argue that research into situated cognition may account for the priority given in medieval psychology to *vis imaginativa* "the ability to imagine" (Carruthers, 2008b, p. 6). One possible account is offered by research into simulation. As we read in Bergen (2007), "through exposure to language in context, language users learn to pair chunks of language like *kick, Mary,* or *John* with perceptual, motor, social, and affective experiences. In subsequent instances of language use, when the original perceptual, motor, social, and affective stimuli are not contextually present, the experience of them is re-created through the activation of neural structures responsible for experiencing them in the first place. This view of meaning is embodied in that meaning depends on an individual having had experiences in their body in the actual world, where they recreate those experiences in response to linguistic input, and use them to produce meaningful linguistic output" (p. 277).

Although the debate over meaning representation remains unresolved, there are good reasons to assume that the situated cognition perspective is not unfounded (de Vega, Glenberg & Graesser, 2008). Our account of relevant empirical findings is inevitably selective and constrained by the question of why imagistic (iconic) representation might be significant for cultural transmission relative to the role of imagery in language. One simulation-based idea of how abstract concepts might be represented appears suited to the nature of the problem. The conception holds that

> abstract concepts are represented by concrete situations…. Likely, a collection of concrete situations that share the abstract concept combine to form a rich representation…. Representation by concrete situations allows these representations to be grounded in sensory-motor simulations. For example, the concept democracy may be represented by *a meeting of parliament, a succession of speeches, votes on proposals during meetings…* and so on. (Pecher, 2013, p. 368)

It is also argued that simulation enables conceptual representations that serve various functions in memory, language, and thought, including text comprehension (Barsalou, 2005). Importantly, if "intentional attempts to combine simulations of conceptual components can produce simulations never experienced" (Barsalou, 2005, p. 625), perceptual input already stored in memory might be reused in novel ways.

Since an abstract concept may be represented by various situations, we conclude that the investigation of abstract concept representation may consist in the examination of situational patterns behind imagery. It is suggested that they derive from cognitive activity which occurs in context. In this way, the notion of situated conceptualisation helps us establish correspondences between Chaucer's usage and the visual culture of his times, the latter taken as the context in which the individual's cognition could operate. Our analysis pertains to the representation of the abstract concept of 'necessity' in the poet's KT and TR, and is aimed to inform the study of language evolution by providing a link between the situated nature of human cognition and linguistic usage.

## 4. Chaucer's imagery in TR and KT: the wheel of Fortune and the amphitheatre

We have indicated that both language and visual arts in the times of Chaucer transmitted knowledge about abstract ideas. Within the scope of this paper it is not possible to offer an in-depth account of the Boethian imagery in KT and TR. Therefore, we zoom in on its selected aspects, choosing for discussion an image of Fortune's wheel in TR and the image of the symbolic cosmos in KT. Both images constitute situational representations of the abstract concept of 'necessity' consonant with Boethius' *Consolation* (see Book II, especially pr. 1; IV, pr. 7; V, especially pr. 1, 3–6), encompassing various conceptions, including ideas of Ptolemy (see *Consolation*, Book II, pr. 7).This syncretism means that each of the images inspired by Boethian visualisations may evoke different aspects of the symbolic tradition.

In the case of TR, one possible way of representing the abstract concept of 'necessity' is its visualisation via the image of the wheel of Fortune. As exemplified in the following excerpt on the trickery of Fortune, Chaucer does not provide a detailed description of the wheel, but leaves many situational aspects for the audiences to infer,

> … From Troilus she gan hire brighte face
> *she began to turn her bright face away from Troilus*
> Awey to writhe, and tok of hym non heede,
> *and took no heed of him*
> But caste hym clene out of his lady grace,
> *but cast him clean out of his lady's grace*
> And on hire whiel she sette up Diomede
> *and on her wheel she set up Diomede* TR, Book IV, ll. 8–11[1]

Various images of the wheel of Fortune, originally a Roman goddess, available in the visual culture of Chaucer's times[2] facilitated thought-sharing within the community, fostering a shared conceptual order. Hence, knowing the image, one might easily infer that whoever ascends the wheel should keep in mind its incessant rotation. This partial image of the wheel of the insidious Fortune allows inferences via completing the situational pattern (see Barsalou, 2005). Thus, for anyone familiar with the cultural common ground it is clear that Troilus who has managed to ascend the wheel must inevitably fall down, making room for his rival, Diomede. However, inferences may go even further. While necessity was associated with God's Providence, hence as such inaccessible to the human mind's eye, "that-which-God-foresees" was orderly implemented in the world of chance via the workings of destiny. The image of Fortune spinning her wheel thus captures the complex interplay between the various forces. Implicit remain values promoted in *Consolation*, including integrity in the face of mutability of fate and calmness in adversity. Insofar as this Boethian image carries important

1 Both quotes from Benson (Ed.) 2008/ 1987.

2 To give but two examples, Bodleian Library MS Douce 298, fol.13v., c.1400, and, for comparison, Royal 18 D II f. 30v, from the time after Chaucer's death. In both the general template remains unchanged.

information concerning the value system shared within the community, it offers a glimpse into the medieval mindset.

While in the Middle Ages the idea of remaking imagery adopted from others, especially writers of the classical world, was widespread, in KT Chaucer was free to reuse ideas from Boccaccio's *Teseida*, including that of the amphitheatre, in the latter possibly based on the Roman Colosseum (Kolve, 1984, pp. 105–106). Insofar as sense-making relied on sharing that medieval cultural common ground, the poet could draw on imagery integral to the symbolic tradition familiar for his audiences. This, however, was not tantamount to creating a carbon copy of the Italian narrative.

Hence, another visualisation of 'necessity' via a concrete situation may be found in the emblematic tournament scene from KT, where once again the poet depicts chance and destiny at work. The cosmic, planetary forces of gods involved are present in a symbolic way. To resolve the love conflict between Arcite and Palamon over Emelye, Theseus orders to construct an amphitheatre, where the decisive tournament is to take place. It is designed according to zodiacal principles such that its arrangement includes temples of the planet-gods located in places corresponding to their planetary positions (see also Kolve, 1984; Tasioulas, 2005). By envisaging the interplay of planetary influences via the zodiacal architecture of the amphitheatre, the poet manages to show how the characters' destinies are fulfilled in the complex interplay of humans' desires at the mercy of the divine will,

… this noble duc gan ryde,
*… this noble duke rode forth,*
Thise two Thebans upon either syde,
*the two Thebans on either side…*
And westward, thurgh the gates under Marte,
*and westward, through the gates under Mars,*
Arcite, and eek the hondred of his parte,
*Arcite, with the hundred of his party,*
With baner reed is entred right anon;
*with red banner entered immediately;*
And in that selve moment Palamon
*and at that very moment Palamon*
Is under Venus, estward in the place,
*under Venus (entered)on the eastern side,*
With baner whyt….
*with white banner….* KT, ll. 2569–2586

Chaucer thus offers a twofold perspective onto the human and the divine worlds. At one level, we see the world of contingencies, where events unfold by the machinery of chance and destiny. At the other, we discern providential motivation governing this world. In the subsequent part of the tale, the horse of the triumphant Arcite is attacked by the fury sent by Saturn to fulfil the request of Venus. While Mars' fighter wins the tournament so as to fulfil Arcite's desire, Saturn restores balance between the quarrelling two gods by setting the final score in favour of Venus' fighter, Palamon, who had asked to win Emelye, rather than the tournament itself. It is Palamon who marries her, because Arcite dies as a result of the accident in which he is unhorsed and terribly wounded. It is thus bitterly ironic that, in the words of Duke Theseus, the ultimate source of cosmic order is the benevolent Jupiter. Insofar as the equilibrium achieved in the tournament necessarily involves death and disruption, Chaucer's is an imagery with disturbing undercurrents, implicitly challenging the consolation of philosophy offered by Boethius.

This conclusion seems to be corroborated not only by the amphitheatre imagery, but also by the visualisations depicting the spherical universe.[3] The sublunary sphere embraced the perishable world of contingencies inhabited by humans, while the superlunary sphere comprised the Moon, the Sun, and planets, constituting the world of necessity. The KT imagery is correspondingly complex, forming a nexus of interconnected images. Chaucer thus combines the imagery of the spherical universe with that of the Theban myth, tingeing it with Boethian overtones. Clearly, the medieval model of the world was syncretic, harmonizing various conceptions. From this amalgamation emerged the conceptual order of Chaucer's community, which pervaded the ways in which the poet and his contemporaries made sense of the world.

It follows that while image may serve as a means of knowing, imagistic (iconic) memory carriers may encapsulate knowledge based on misguided assumptions. Yet, imagery can be tailored to the specific sociocultural situatedness of the individual, as shown in KT and TR. Even incomplete images may offer quick access to situational patterns from symbolic tradition, which enables inference-making. Imagery shared via external representations may foster the continuity of the collective conceptual order, safeguarding imagery conveyed via language.

## 5. Concluding Remarks

We have argued that human cognition is situated, which, on our account, entails the use of external representations. Hence, language and artefacts of visual culture have been shown as complementary memory carriers encapsulating the

3 As in Harley MS 3647, f. 22v (Sacrobosco's *De sphaera mundi*, circa 1230).

community's conceptual knowledge. Insofar as image can serve as a means of access to that knowledge, the medieval visualisations of abstract concepts we have examined appear to reveal the significance of the state of cultural evolution achieved by the community for the representation of abstract ideas.

As a visual poet, Chaucer seems to have used shared imagery integral to the cultural common ground of his community. Thus, in KT and TR, the poet provided relevant iconographic details easily recognisable by and/ or understandable for his contemporaries. The significance of images might in part have derived from the symbolic tradition inherited from the past. However, the character of memory carriers available to the medieval community was not unimportant either. This stems from the fact that the human mind was believed to serve as the best repository of cumulative memory. That is, the imagistic (iconic) nature of representations of 'necessity' in KT and TR might follow from the medieval views on the human mind.

We have shown how the images served as gateways to the contemporary symbolic tradition, and underscored the situational nature of the imagery associating contextually relevant cues (iconographic details). For this reason, we have highlighted the situational character of the image of Fortune's wheel, suggesting that this compact way of encoding information might be seen as one effective means of transmitting cumulative knowledge. The amphitheatre scene offers an equally compact image filled with situational cues that allow inference via pattern completion based on participation in the contemporary cultural common ground. Chaucer's sociocultural situatedness may thus be seen as motivating the situational representations of 'necessity', providing ways to visualise it in novel ways by reusing perceptual input traceable in language and visual culture that derived from the experience of past generations.

Overall, insofar as human knowledge is situated, verbal and visual channels of cultural transmission may overlap in the coherent whole of the multimodal image. It seems reasonable to conclude that due to the capacity for combining linguistic symbols with stored perceptual information humans may transcend the constraints of the present using language and visual culture resources as memory carriers to share thoughts.

## References

*Barsalou, L.* (2005). *Situated Conceptualization.* In H. Cohen & C. Lefebvre (Eds.), *Handbook of categorization in cognitive science* (pp. 619–650). St. Louis: Elsevier.

*Benson, L. (Ed.).* (2008) [1987]. *The Riverside Chaucer* (3rd edition). Oxford: OUP.

*Bergen, B.* (2007). Experimental methods for simulation semantics. In M. Gonzalez Marquez, I. Mittelberg, S. Coulson, & M. J. Spivey (Eds.), *Methods in Cognitive Linguistics* (pp. 277–299). Ithaca.

*Bernárdez, E.* (2008). Collective cognition and individual activity: Variation, language and culture. In R. Frank, R. Dirven, T. Ziemke, & E. Bernárde (Eds.), *Body, Language and Mind. Volume 2: Sociocultural Situatedness* (pp. 137–166). Berlin-New York: Mouton de Gruyter.

*Boroditsky, L. & Prinz, J.* (2008). What thoughts are made of. In G. R. Semin & E. R. Smith (Eds.), *Embodied Grounding: Social, Cognitive, Affective, and Neuroscientific Approaches* (pp. 98–115). Cambridge: CUP.

*Brown, P.* (2011). *Authors in context. Geoffrey Chaucer.* Oxford: OUP.

*Carruthers, M.* (2008a). *The Book of Memory.* Second edition. Cambridge: CUP.

*Carruthers, M.* (2008b). Mechanisms for the transmission of culture: the role of 'place' in the arts of memory. In L. Hollengreen (Ed.), *Translatio, the Transmission of Culture in the Middle Ages* (pp. 1–26). Arizona Studies in the Middle Ages and Renaissance (Turnhout: Brepols).

*Carruthers, M.* (2013). Virtue, intention and the mind's eye in Chaucer's Troilus and Criseyde. In C. Brewer & B. Windeatt (Eds.), *Traditions and Innovations in the Study of Middle English Literature* (pp. 73–87). Woodbridge: Boydell and Brewer.

*de Vega, M., Graesser, A., & Glenberg, A.* (2008). Reflecting on the debate. In M. de Vega, A. Glenberg, & A. Graesser (Eds.), *Symbols and embodiment: Debates on meaning and cognition* (pp. 397–440). Oxford, U.K.: OUP.

*Evans, V., & Green, M.* (2006). *Cognitive Linguistics: An Introduction.* Edinburgh: EUP.

*Frank, R.* (2008). Introduction. In R. Frank, R. Dirven, T. Ziemke, & E. Bernárdez (Eds.), *Body, Language and Mind. Volume 2: Sociocultural Situatedness* (pp.1–18). Berlin-New York: Mouton de Gruyter.

*Golka, M.* (2009). *Pamięć społeczna i jej implanty* [Social memory and fake memory carriers], Warszawa: Scholar.

*Hałas, E.* (2012). Przeszłość i przyszła teraźniejszość: refleksyjna pamięć kulturowa [Past and future present: reflective cultural memory]. In E. Hałas (Ed.), *Kultura jako pamięć* [Culture as memory] (pp. 153–176). Kraków: Nomos.

*Kirsh, D.* (2010). Thinking with External Representations. *AI and Society*. Springer: London, 25: 441–454.

*Kolve, V.* (1984). *Chaucer and the Imagery of Narrative: The First Five Canterbury Tales.* Stanford University Press.

*Mann, J.* (2003). Chance and destiny in *Troilus and Criseyde* and the *Knight's Tale*. *In J. Mann & P. Boitani (Eds.), The Cambridge Companion to Chaucer (*2nd edition) (pp. 93–111). Cambridge: CUP.

*Pecher, D.* (2013). The Perceptual Representation of Mental Categories. In D. Reisberg (Ed.), *The Oxford Handbook of Cognitive Psychology* (pp. 358–374). New York: OUP.

*Ripa, C.* (2013). *Ikonologia* [Iconologia]. Kraków: Universitas.

*Robbins, P. & Aydede, M.* (2009). A short primer of situated cognition. In P. Robbins & M. Aydede (Eds.), *The Cambridge Handbook of Situated Cognition* (pp. 3–10). Cambridge: CUP.

*Smith, E. R., & Collins, E. C.* (2010). Situated Cognition. In B. Mesquita, L. Feldman Barrett, & E. Smith (Eds.), *The mind in context* (pp. 126–145). New York: Guilford Press.

*Smith, E.R. & Conrey, F.R.* (2009). The Social Context of Cognition. In P. Robbins & M. Aydede (Eds.), *The Cambridge Handbook of Situated Cognition* (pp. 454–466). Cambridge: CUP.

*Smith, L. & Colunga, E.* (2012). Developing Categories and Concepts. In M. Spivey, K. McRae, & M. F. Joanisse (Eds.), *The Cambridge Handbook of Psycholinguistics* (pp. 283–309). Cambridge: CUP.

*Tasioulas, J.* (2005). Science. In S. Ellis (Ed.), *Chaucer.* An *Oxford Guide* (pp. 174–189). Oxford: OUP.

*Tomasello, M.* (2014). *A Natural History of Human Thinking*. Cambridge, Massachusetts-London, England: Harvard University Press.

*Zalewska, A.* (2014). Palimpsest. In M. Saryusz-Wolska & R. Traba (Eds.), *Modi. memorandi. Leksykon kultury pamięci* [*Modi memorandi.* Lexicon of the culture of memory] (pp. 318–322). Warszawa: Scholar.

Martin Hinton
*University of Łódź*

# Alternatives to intuition in linguistics research

**Abstract:** Much research in linguistics relies upon intuitions: either those of the researcher himself or those of speakers of the language under consideration. There are a number of reasons to think that this reliance may not always be the best way to proceed; and there are unsolved questions as to exactly whose intuitions make the most reliable data; what those intuitions actually are; and what they should be about. I begin with a short review of these debates and propose three possible alternatives to researcher intuitions: a broader based survey, a language corpus search and internet search engines. Methodological issues with all three are discussed with the help of a small scale practical investigation into two suspect intuitions provided by published linguists.

Keywords: intuition, corpus linguistics, methodology, surveys, internet search.

## 1. Introduction

The use of intuitions is perhaps the one thing which unites all schools of linguistics. If linguistics is interested in what speakers know about language, rather than what they actually produce during language use, then the intuitions of those speakers seem to provide the most direct line available to that knowledge. It is certainly a common practice for researchers to list examples of sentences which they consider acceptable and others, usually adorned with an asterisk, which are beyond the pale. These statements of acceptability are not usually supported by anything more than the author's own knowledge of the language. Nothing more appears to be necessary: since the statement is not one about actual use, no examples from the 'real world' are thought relevant, and since the writer is, by definition, an expert, there is no need to consult with other speakers, particularly less expert ones. If the researcher feels with great certainty that an expression or construction is not permissible, then that is apparently enough to use its impossibility as a building block in a wider theory.

There are, however, several questions which have been raised and not satisfactorily answered about both language intuitions themselves and their legitimacy as evidence for linguistic theories. These range from the fundamental, such as, asking what an intuition is, to the more practical matter of how best to elicit one. The lack of reflection on these issues has, according to some writers (Gibbs, 2007; Itkonen,

1981) given linguists something of a bad reputation with researchers in other disciplines and frequently led to calls for a more empirical basis for linguistic work.

The first section of this paper will briefly rehearse the discussion of some of these issues, paying particular attention to the arguments set out stridently by Wasow and Arnold (2004) and the work of Michael Devitt and those who have responded to him. Much of the criticism of the use by authors of their own intuitions, however, has paid little heed to the necessity to provide other methods by which linguists may proceed in their investigations. The subsequent section, therefore, will consider three possibilities: surveying the intuitions of larger groups, searching language corpora, and performing internet searches. Each of these is discussed both from a theoretical point of view and in terms of the appropriate methodology. In order to better understand the methodological issues involved, and to understand what sort of results each procedure might deliver, the third section of the paper describes a small scale study into two intuitive acceptability claims confidently made in the literature, but which I consider at least suspect. The two claims are tested by all three methods and the results presented and discussed.

The concluding section considers the relevance of results generated by these methods to the linguistics project as a whole and whether or not they have anything to tell us about intuitions in general.

## 2. Intuitions in linguistics

There is no doubt that intuitions are extremely useful in linguistics. A researcher may collect his own intuitions very quickly, at no cost, and use them to assess any theory before him. And, when the project of linguistics is conceived as the study of what speakers know when they know a language, there seems little reason to doubt that the researcher himself knows it as well as anyone. Certainly, 'many generative grammarians appear to regard primary intuitions as more direct evidence of linguistic competence than other types of data' (Wasow & Arnold, 2005 p. 1484). Those who have expressed concern have not generally been opposed to the use of intuitions per se, rather to the way in which they are used. Esa Itkonen writes scathingly of a 'lack of methodological self-understanding' (Itkonen, 1981 p. 130) and Wasow and Arnold go on to suggest, quite reasonably, that: 'Since knowledge of language is not directly observable, linguists should use every type of evidence available' (2005 p. 1484), for them, it is not the use of intuitions which is the problem, but the disregard for anything else.

Perhaps the most important argument against the use of researcher intuitions is the well-known level of variation across speakers. This is not simply a question of confusion or hesitancy: 'what one speaker finds unequivocally well-formed another speaker may find unequivocally ill-formed' (Wasow & Arnold, 2005 p. 1482). What reflects so badly on practitioners in the field is that linguists know this and yet still 'typically provide judgements as if they held for all speakers, without checking to see how variable they are' (Wasow & Arnold, 2004 p. 1483). Intuitions come to us with great certainty which appears to make them safe, but in that certainty lies the danger: we do not feel the need to provide other evidence and are dismissive of conflicting views.

## 3. The alternatives

In this section I shall briefly describe three possibilities for linguists who either wish to find confirmation of their own intuitions or would rather avoid making any personal judgements and instead focus at once on the language community, whose ultimate collective agreement is what makes language use possible.

### 3.1 Surveys

Perhaps the most obvious way to seek confirmation of an intuition is to ask others if they share it: if most respondents agree that the language works in a particular way, even if there is still some variability of opinion, it would seem reasonable to believe that that is, in fact, the case, and should be reflected in any grammatical theory. Two questions, however, have dominated debate on this topic: who to ask, and what to ask them.

Michael Devitt has argued that linguists make better subjects. He suggests that: 'As a result of their incessant observation of language, guided by a good theory, linguists are reliable indicators of syntactic reality' (Devitt, 2006 p. 111). The reason for this is that although he calls intuitions 'unreflective judgements' he also believes them to be theory-laden: 'so that those who know more and better theory issue more reliable judgements' (Culbertson & Gross, 2009 p. 2). Aside from leaving us to decide what constitutes 'good theory', Devitt seems oblivious to the evident circularity of this stance. He seems to be suggesting that linguists can use their knowledge of theory to check whether or not their theory fits their own knowledge. If intuitions are indeed loaded with theory, then nobody who has been exposed to the theory can be considered an impartial judge of it.

Culbertson and Gross (2009) conducted their own experiment which showed that linguists are no more reliable as judges than others who have had some contact with cognitive science. Their results, however, are open to some doubt (see Devitt, 2010). One reason for that is the second problem: how should intuitions be elicited? Much hinges on the understanding of the concept of 'acceptable'. Devitt believes this almost always means 'grammatical', others disagree. Culbertson and Gross told their subjects: 'A sentence sounds good if you think you would or could say it under appropriate circumstances' (2009 p. 13), which doesn't seem to me to refer to either grammaticality or acceptability.

### 3.2 Internet searches

Clearly, internet search engines provide a simple way to access a large amount of linguistic material. The methodology should be relatively simple but there are a number of important considerations. Firstly, much will depend on what exactly is searched for and a small difference in search term could be vital. Secondly, website designers work hard to skew the results in their favour so the first hits may not be the best and the sheer number of results usually returned makes it impossible to investigate all of them and makes statistical analysis impossible. This is pointed out by Roland Bluhm (2013), who is an advocate of the use of corpora, and also notes that since many internet users of English are not native speakers the existence of a structure may not mean that it is acceptable to those with better knowledge of the language. Most importantly, the presence or absence of a term on the web proves little in itself.

### 3.3 Corpora

Corpora have become a fundamental part of applied linguistics research and are beginning to be advocated for use in general linguistics as well. Bluhm provides a full discussion of appropriate methods and suggests four advantages of corpora. They:

> 'provide data on the basis of which hypotheses can be formulated [...] all of these data are, by and large, unfiltered [...] the contexts in which the queried expressions are found give insights into the variety of real-life situations in which the phenomenon referred to by the concept occurs [and] the contexts often provide excellent raw material for thought experiments with regard to the concept *and* the phenomenon in question.' (Bluhm, 2013 p. 12)

He accepts that not all questions of linguistics can be studied in this way, but maintains that the continued reliance on intuitions and intuitions alone is untenable.

## 4. The study

It isn't difficult to find dubious example of supposedly good and bad sentences in linguistics papers: the very nature of intuitions means that no evidence is offered in their support, and researchers, obviously, write mainly about that which is open to discussion. The two which I have chosen to investigate further appealed to me particularly because they did seem to concern things which the authors thought ordinary speakers would say, and might, therefore, lend themselves more obviously to empirical study. The first concerns the possibility of a type of construction and comes from a study in pragmatics into negative polarity licensing by Jay David Atlas (2007). He cites the example sentence: 'I know no one distinct from Hillary trusts Bill, but does only Hillary trust Bill?' from another paper (Horn, 2002) where it was considered impossible. I would certainly agree with Horn on that, but Atlas doesn't. Two things make this example interesting: firstly, despite being fully aware that another researcher has intuitions disallowing this structure, Atlas feels no compunction in asserting that his intuition is correct, without offering any evidence or support. Secondly, and most usefully for this study, he informs the reader that such an expression would be familiar to 'the English middle classes' (Atlas, 2007 p. 21). This claim, then, is particularly susceptible to empirical investigation: one must simply ask the English middle classes, or at least their representatives, what they think.

The second dubious example, of a different nature, is found in a cognitive linguistics textbook by Vyvyan Evans and Melanie Green and purports to show how encyclopedic knowledge allows apparently contradictory sentences to be understood. The example is: 'Watch out Jane, your husband's a right bachelor!' (Evans & Green, 2006 p. 161). This immediately struck me as odd, but the authors go on to explain in the next paragraph that it makes sense due to our 'cultural stereotype of bachelors, which represents them as sexual predators'. This example, then, is based upon an intuition about the connotations of certain words. My intuition is very different, which led me to Roget's Thesaurus (1988). 'Bachelor', in the sense of unmarried man, is listed under one heading: 'Celibacy'. If the cultural stereotype of a bachelor is indeed of a sexual predator, it must be a fairly recent shift and that is something which can be investigated empirically.

Both of these acceptability assertions were subjected to three tests as outlined in the section above.

## 4.1 Test 1 – The survey

The survey was conducted on a small-scale using 10 participants. All 10 were educated, middle-class Britons: 7 Englishmen and 3 Scots. They completed the survey individually in my presence and were free to ask questions about the survey itself, though not about the sentences under investigation. The survey consisted of 10 sentences and is reproduced in full in the table below, together with the instruction question. The numbers in each column represent the number of responses placed there by the participants. The asterisks are my own intuitions.

Would you consider the following sentences acceptable English or not?

*Table 1: The survey*

| | **Acceptable** | **Hard to say** | **Unacceptable** |
|---|---|---|---|
| 1. I love only you, but I'm not sure that I love you. | 3 | 3 | 4 * |
| 2. No-one except John supports Bill, but does John support Bill? | 5 | 1 | 4 * |
| 3. I love only you, nobody else. | 10 * | 0 | 0 |
| 4. No-one other than John voted for Bill, but did John vote for Bill? | 4 * | 3 | 3 |
| 5. No-one other than you loves me. | 5 * | 1 | 4 |
| 6. I know no-one distinct from John trusts Bill, but does only John trust Bill? | 1 | 2 | 7 * |
| 7. I know John trusts Bill, but does only John trust Bill? | 8 * | 1 | 1 |
| 8. Watch out Jane, your husband's a right bachelor! | 2 | 1 | 7 * |
| 9. Even though he's married, he's still a bit of a bachelor. | 8 * | 0 | 2 |
| 10. No-one but me loves Jane, but do only I love Jane? | 2 | 0 | 8 * |

It will be noted that the Atlas example has been changed to remove the Hillary/Bill connection which, with its political associations, I believed would be a distraction. A number of observations can be made at once: sentences 6 and 8, the focus of the survey, were both rejected by 7 out of 10 respondents. In the case of sentence 6 which was supposed to be common currency among the English, removing the Scots from the results would leave columns of 1 0 6. Interestingly, there is general support for sentence 9 which repeats the married-man/bachelor inconsistency but without the warning. This suggests to me that while Evans and Green are right about how we understand words, they simply have a different stereotype of bachelors from those who were surveyed.

The fundamental purpose of this survey, however, was to assess the method itself. By using a small group and sitting with each individual it was possible to collect spontaneous feedback. As many as 8 of the 10 participants asked what was meant by 'acceptable'. It is clear that they could see there might be different types of acceptability, and several asked if it meant simply 'grammatical'. One participant told me that, for him, anything which conveyed a message was acceptable, and rejected only sentences 8 and 10.

It is clear too that there is a good deal of inconsistency. Only sentence 3 attracted a unanimous assessment. This was a surprise as items 5 and 7 were also included as obvious and uncontroversial examples and it is a mystery to me how 5 educated English native speakers could doubt the validity of sentence 5. That, of course, is the point. Seeing one's own solid intuitions doubted and denied is a sobering lesson for any professional linguist. Two participants marked the 'Hard to say' box three times, and only four not all. This suggests that native speakers do not necessarily know whether a particular form is acceptable or not, a finding which goes to the very core of our understanding of what it is to know a language. Clearly, it does not mean, even for articulate native speakers, to be able to judge all language items as valid in their language or not.

The results might be looked at differently, though. If we accept, somewhat arbitrarily, a 70% agreement rate as decisive, then the acceptability of six of the sentences has been established, and the other four may be considered difficult cases. Discovering the reasons for this difficulty would then become an interesting line of research. Wasow and Arnold refer to 'secondary' intuitions, that is, intuitions about why our intuition tells us something is acceptable or not, which are also common in the literature and all of the issues surrounding first order ones apply to them too.

## 4.2 Test 2 – The internet

The internet search was conducted using Google.com. The choice of search query is obviously of the utmost importance, so it seemed sensible to try a variety of ways of phrasing the item under investigation, where possible. The husband/bachelor contrast appeared readily suitable for this type of search. The entire sentence (no. 8 above) achieved 0 exact matches, and the first 100 results yielded no uses of 'bachelor' similar to that in the suspect phrase. The words 'husband bachelor' produced many results (in the millions, which is a problem for this type of study) but the first 100 again showed nothing related to the Evans and Green example. 'Bachelor husband', however, was found to have been the title of a novel and to be a phrase in use on social media. Unfortunately for Evans and Green, the definitions offered were not largely concerned with promiscuity. Facebook (see below for page references) contains a full definition: 'BACHELOR HUSBAND Does things on his own without consulting wife. Hangs out a lot with friends more than wife.

Not serious about marriage (sic) life.' A letter to the Irish Independent newspaper (2003) describes one as 'a good-time guy with a short attention span. He has no interest in the house, the children, or me.' A bachelor husband, in short, is not one who necessarily chases other women, but one who is not interested in married life and continues to live as though he were single.

A search for the Atlas phrase proved far more problematic. The whole sentence returned two articles by Atlas and another one quoting him. The remainder of the first 100 results, unsurprisingly, produced nothing of interest to the study. The phrase 'no one distinct from' had only one exact match: a citation of Horn discussing the phrase. 100 results yielded nothing remotely similar. Since it was hardly worth searching for 'only' the test ended there.

### 4.3 Test 3 – Language Corpora

For the corpora search I used the Brigham Young University portal (Corpus. BYU.edu) to access the British National Corpus (BNC), the Corpus of Contemporary American English (COCA) and the very up-to-date Corpus of Global Web-Based English (GloWbe). The husband/bachelor collocation with a possible separation of 9 words either side delivered 4 matches on BNC, 9 on COCA and 22 on GloWbe. None of the matches was relevant to the Evans and Green sentence, although interesting expressions such as 'bachelor bed' and 'bachelor TV' were discovered, further suggesting that when bachelor and husband collocate, it is to refer to elements of the husband's former life, not, or at least not specifically, to his sexual practices.

The Atlas sentence was more difficult to search for. Interestingly, all three corpora delivered zero results for the phrase 'no one distinct from', and even 'no one distinct', under all spellings of 'no one'. The BNC had 'distinct from' and 'only' collocated within a 9 word separation either side 15 times, the GloWbe 83 times and COCA not at all. None of the uses recorded was similar to the Atlas sentence.

## 5. Discussion and conclusion

First, and most importantly, none of the three tests lends any support to the authors whose intuitions I doubted. In the first test a clear majority, 7/10 for both sentences, found their examples unacceptable, and neither the internet search nor the corpora search turned up any actual uses of their constructions, or indeed, anything similar. The question for linguists, however, is do such result matter? The difficulties with wider intuition surveys were discussed above and the lack of evidence in the corpora and internet searches is clearly not evidence of the unacceptability of a sentence, nor even of its general lack of use, since corpora are obviously limited in their scope. It does seem, however, as though these results ought to give researchers pause for thought. The survey shows that there is a large grey area for many speakers between good and bad sentences; the searches show that phrases which may seem quite reasonable to the linguist are not in fact used by speakers of the language. Theoretical linguists dealing only with what a particular grammar makes possible may feel able to ignore other evidence, but the authors of the two examples under examination in these tests both made clear references to their normal use: Atlas suggested a particular group would accept the construction: they didn't, Evans and Green claimed a word was attached to a particular stereotype, an empirical claim for which no evidence has been found.

## References

*Atlas, J.D.* (2007). On a pragmatic explanation of negative polarity licensing. In N. Burton-Roberts (Ed.) *Pragmatics* (pp. 10–23). Basingstoke: Palgrave Macmillan.

*Bluhm, R.* (2013). Don't ask, look! Linguistic Corpora in Philosophical Analyses. In M. Hoeltje, T. Spitzley & W. Spohn (Eds) *Was dürfen wir glauben? Was sollen wir tun?* DuEPublico.

*Culbertson, J. & Gross, S.* (2009). Are Linguists Better Subjects? *British Journal for the Philosophy of Science, 60*, 721–36.

*Devitt, M.* (2006). Intuitions in Linguistics. *British Journal for the Philosophy of Science 57*, 481–513.

*Devitt, M.* (2010). Linguistic Intuitions Revisited. *British Journal for the Philosophy of Science 61*, 833–865.

*Evans, V. & Green, M.* (2006). *Cognitive Linguistics – An Introduction*. Edinburgh: Edinburgh University Press.

*Horn, L.* (2002). Assertoric Inertia and NPI Licensing. *CLS 38, Part 2. Chigaco Linguistics Society*, 55–82.

*Itkonen, E.* (1981). The Concept of Linguistic Intuition. In F. Colmas (Ed.) *A Festschrift for the Native Speaker* (pp. 127–140). The Hague: Mouton.

My Bachelor Husband Repulses Me. (2003, November 30) Retrieved from: www.independent.ie/lifestyle/my-bachelor-husband-repulses-me-26237593.

*Wasow, T. & Arnold, J.* (2005). Intuitions in linguistic argumentation. *Lingua 115*, 1481–1496.

Where Do Broken Hearts Go. (2013, May 17). Types of husbands. Retrieved from: www.facebook.com/nasasirapages2/posts/487296884673302.

Anna Bączkowska
*Kazimierz Wielki University, Poland*

# Compliments in film subtitles: a pragmatic and cognitive study of translations from English into Polish

**Abstract:** The primary goal of this paper is to present a novel approach to the analysis of compliments in subtitles. The methodology advocated in the study conflates a pragmatic perspective on compliments with tools deriving from cognitive linguistics known as construal operations. The applicability of selected tools used in cognitive linguistic analysis to audiovisual translation is demonstrated on the basis of several scenes gleaned from a romantic comedy. The study demonstrates that construal operations may enrich the traditional analysis of subtitles by revealing details concerning the possible conceptualisations of compliments in different languages.

Keywords: Compliments, construal operations, subtitles, film discourse, cognitive linguistics.

## 1. Introduction

The primary goal of this paper is to present an alternative approach to the analysis of compliments in subtitles. The proposed research tools, specifically the so-called construal operations, which are borrowed from cognitive linguistics (CL), are believed to yield an insightful description of different types of compliments in filmic discourse. Audiovisual Translation, and subtitles in particular, may benefit from the application of construal operations as such an analysis is detailed and thorough, and it may complement a discussion on subtitling strategies. Several instances of compliments will be retrieved from the original English soundtrack of a romantic comedy, and they will be compared with their renditions gleaned from the Polish subtitled version. The sample scenes will be examined for illustrative purposes in order to demonstrate that by enriching a pragmatic analysis with a cognitive approach to film translation, a finely-toothed description may be offered.

## 2. Theoretical assumptions

Complimenting has been the topic of research for years now. It has attracted a number of scholars who approach this issue from different perspectives and who study sundry aspects of compliments. There is a plethora of studies revolving around their linguistic structure, compliment responses as well as the types and functions of compliments. Of a number of functions, the most widely quoted involve: the creation or reaffirming of the feeling of solidarity and common ground shared by interlocutors, establishing rapport, bolstering 'face' (Goffman, 1967), establishing social cohesiveness or sheer expression of admiration. Compliments may centre on a number of topics, such as appearance, personal belongings, personal traits, career, skills, socio-economic status, etc. They tend to be rather formulaic and repetitive, and have a phatic function. The structures typical of compliments were mainly expounded in the flurry of writings in the eighties by Manes and Wolfson (1980), Wolfson and Manes (1980), Wolfson (1984) and Holmes (1986).

The definition of a compliment most oft-cited is probably one proposed by Holmes (1986: 485, 1995: 117). She maintains that a compliment is a type of speech act which "explicitly or implicitly attributes credit to someone other than the speaker" and which is "positively valued by the speaker and the hearer". Compliments are generally divided into direct and indirect (Kerbrat-Orecchioni, 1987, Y. Yuan, 2002), as well as explicit and implicit (Kerbrat-Orecchioni, 1987, Boyle, 2000); occasionally, lesser known classes of compliments are proposed, e.g. solicited vs. unsolicited (Bruti, 2009), or personal (which include explicit, indirect and implicit), ceremonious, season and free gift compliments (Jucker, 2009). Direct compliments take place when the speaker addresses the compliment recipient straightforwardly, leaving no room for doubts or making wild guesses (*You are beautiful*; *Well done*), while indirect compliments require that a third party be involved, either by comparing to another person's qualities or by quoting another person's words (*You look like your beautiful mother, Your father says your English is perfect*). Explicit compliments, on the other hand, are similar to direct compliments as they praise the compliment receiver openly and straightforwardly with a complimenting assertion (*I love your dress*), implicit compliments being reserved for more camouflaged cases (*Your husband has a very good taste* while complimenting his wife, *I wish I could sing like you do*, or *–What a polite child! – Thank you, we do our best*).

Because of the diversity of typologies of compliments and, all too often, fuzzy boundaries or overlapping categories, a simplified typology has been employed in the analysis which follows, wherein three main labels are used: direct, indirect and implicit compliments. While the direct compliment is self-explanatory and converges with the definitional profiles of the previous proposals, the last two cases may appear vague. In our understanding of these terms in this paper, a direct compliment addresses the hearer outspokenly (*You are/look beautiful*), while an indirect compliment entails a comparison between compliment addressee and the third party (e.g. through the use of 'like' or 'as if'). Finally, implicit compliment is used of contexts wherein on the surface there is no direct noticeable relation between the compliment-giver and compliment-receiver, via, for example reference to third party (quoting others, the use of 'as if' or 'like' structure, etc.), yet at the same time praise is sensed and it is inferable. This is because for the stakeholders the hiatus between the known and the new information is small enough to retrieve the elided message from indexicals and situational clues. Owing to the intended recoverability of pragmatic ellipsis, the compliment recipient and the viewer may successfully resort to felicitous guesses. Implicit compliments would thus be associated with greater distance existing between the co-participants compared with indirect compliments. Here are examples retrieved from our data:

> Indirect: You are like a genius, you know? (English, Scene 3, subtitle 132).
> Implicit: Beautiful apartment and the view. (complimenting on Nick's taste; back-translation from Polish, Scene 27, subtitle 962).

Generally speaking, a linguistic analysis of complimenting lies within the scope of pragmatics; however, pragmatic topics have rarely been under inspection in audiovisual translation. Such notions as features of spoken discourse and politeness have been only recently, but still fairly sporadically, an object of research into the nature of filmic discourse. As far as the spoken discourse is concerned, the topics revolve around the following issues: conversation analysis applied to subtitling (Tomaszkiewicz, 2006, 2009), register and spoken language in Portuguese subtitles (Rosa, 2001) and Italian dubbing (Pavesi, 2008, 2009), the nature of colloquial language, greetings and leave-takings in Italian dubbed films (Bonsignori, Bruti & Masi, 2011), vocatives in Italian subtitles (Bruti and Perego, 2008), and interjections in Catalan (Matamala and Lorente, 2008). On the other hand, the issue of subtitling politeness has been explored by Hatim and Mason (1996), Gricean cooperation rules in Norwegian subtitles have been examined by Skuggevik (2009), and in Polish subtitles by Bączkowska (2012b), while rendering politeness in Chinese subtitles has been studied by X. Yuan (2012). More importantly for the present study, an increasing interest may be observed in the following topics: formulaicness of compliments in subtitles (Freddi, 2008), direct, indirect and implicit compliments in Italian subtitles (Bruti, 2006, 2009), multimodal aspects of Polish subtitles encoding compliments (Bączkowska, 2011, 2012a), and compliments in films in the context of language teaching (Rose, 2001)[1].

Aside studies into discourse and compliments in subtitles mentioned above, more compellingly, it is only very recently that another approach to subtitles, seen through the lens of cognitive linguistics (CL), has seen the light of day (Caimi 2012, Deckert 2013). Being still in gestation, this method advocates the applicability of the so-called construal operations to the analysis of subtitles. And indeed, this method seems to be more than manageable. CL is known for crossing borders, drawing on neighbouring disciplines of science and relying on a theoretical-methodological framework promoting subjective, experience-based and embodied explication of meaning. This view is well portrayed by construal operations as presented in what follows. Not many studies combining the two perspectives have been offered so far. I have come across only two publications on this topic: a paper by the linguist Annamaria Caimi (2012) investigating the applicability of a cognitive-based approach to subtitles in the context of subtitler training, and Mikołaj Deckert's (2013) doctoral dissertation on construal operations applied to subtitles in documentary films. To the best of my knowledge, the applicability of the cognitive linguistic paradigm to subtitles is still rare, and specifically construal operations have not been widely employed so far. It is thus the primary objective of the paper to present several analyses illustrating the applicability of construal operations to interlingual subtitles.

## 3. Data analysis

### 3.1 Method of investigation

The method of investigation proposed in this paper, overarching pragmatics and cognitive linguistics, invites a fuller picture to emerge. The two-pronged approach mapped onto our data is believed to enhance the analysis of interlingual subtitles by projecting a comprehensive perspective on the one hand, and by offering an impregnated and precise delineation of meaning on the other, thus yielding more profound research outcome compared with a single-track approach.

The analysis discussed in what follows is empirical in nature as it is a data-driven study, with the material culled out from a corpus of film discourse. Data retrieved from films represent fictional language. According to Jucker (2009: 1615), fictional language belongs in a wider group classified as naturally occurring data,

1 Compliments in Polish were discussed by Lewandowska-Tomaszczyk (1989), Herbert (1991) and Jaworski (1995).

i.e. actual language rather than data elicited specifically for the purpose of a study. He further contends that research based on naturally occurring data rests on observation (rather than introspection or a scientific experiment) and it is explored in line with the so-called field study method (as opposed to armchair method or laboratory method). Following this commitment to research, a method based on naturally occurring material–including fictional language–exemplifies empirical studies.

Film discourse illustrates the so-called telecinematic discourse and it represents a special type of natural data. Several facets of film discourse have been distinguished by experts in the field (see Piazza, Bednarek & Rossi, 2011). First and foremost, telecinematic discourse is believed to re-present and re-create real life discourse (Piazza et al., 2011). However, compared with authentic language, it is more coherent and cohesive, less spontaneous, has higher lexical density and there are fewer interruptions in filmic language than in natural discourse (Pavesi, 2009, Bruti 2009, Rossi, 2011). Film dialogues in general, and subtitles in particular, are expected to shun verbosity. In Italian, for example, the typical length of a naturally occurring utterance is six words while in films it is four words; the length of turns, on the other hand, is c. thirteen words in natural discourse as opposed to c. eight words in filmic discourse (Rossi, 2011: 30). Reduction in subtitles results from subtitling principles, the so-called space and time constraints, that typically prescribe two lines per subtitle (each containing only about 30 characters) which can be displayed within no more than six seconds (Díaz Cintas and Remael 2007). Moreover, subtitles may be substantially reduced if non-verbal signs (notably those coded by pictorial elements) are taken into consideration. The interplay between verbal, aural and visual signs is of primary importance as all these signals are integrated in order to create a meaningful and cohesive text (Piazza et al. 2011); this approach is known in subtitling as multimodality (cf. Delabastita, 1989, Taylor, 2003, 2004, 2009, Gambier, 2006, Bączkowska, 2012b, Bączkowska & Kieś, 2012). It must be borne in mind that subtitles depart from authentic language even more than scripted dialogues due to a stylistic modification, known as diamesic shift, prescribed by experts in the field (Gottlieb, 2008). In a nutshell, subtitles are expected to represent language which is less spontaneous than the scripted dialogues, yet, at the same time, they are not supposed to be as formal as written discourse. Taken together, the scripted dialogues represent authentic conversations with some limitations and as such cannot be readily generalised to natural data (Rossi, 2011); still, film discourse is claimed to correspond to authentic language with high credibility (Rose 2001, Alvarez-Pereyre 2011).

Returning to our study, both the original soundtrack and the professional translation thereof constitute the data for our investigation, and the proposed method is based on construal operations. By employing construal operations it is possible to yield more insightful and fruitful descriptions than by resorting solely to pragmatic descriptions (typology of compliments and their syntactic structures), which has been demonstrated by the quantitative analysis presented in this paper: the results of our analysis show little or hardly any differences between the original soundtrack and the subtitled version when construal operations are ignored, and noticeable differences when they are taken into consideration.

## 3.2 Research results

The present section analyses compliments retrieved from a romantic comedy "What Women Want", directed by Nancy Meyers, first aired in 2000, starring Mel Gibson (Nick Marshall) and Helen Hunt (Darcy McGuire). As the main character is a good illustration of a typical American womaniser, I expected to find a number of compliments in the film, and that was the reason for selecting this specific romantic comedy for my study.

The structure of the analysis is as follows. A brief report on the quantitative data will be presented prior to a qualitative examination of selected scenes which represent three types of construal operations. The qualitative investigation is by no means meant to be exhaustive, rather it is meant to be illustrative of how construal operations can contribute to the overall meaning emergence and, in the wake of these considerations, how construal operations can find place in audiovisual translation.

### *3.2.1 Quantitative data – types of compliments*

As the film depicts a character who is a womaniser, and complimenting appears to be an intrinsic strategy of gaining women's hearts, the number of compliments selected for the analysis is high and it reaches as many as fifty six cases in the English script. The subtitled version in turn contains fifty cases of compliments. As for the typology of compliments, there are some noticeable quantitative differences between the English original soundtrack and the Polish subtitles, yet they are not significant. The proportion between direct and the two subtypes of implied compliments (indirect and implicit) spreads almost equally. Qualitatively, the differences are more significant yet the taxonomy of compliments, while valid for a pragmatic analysis, does not manifest *how* they vary in terms of other parameters; this will be discussed in the subsequent section. The table below visualises the total number of compliments occurring in the two versions and the percentages for direct (D), indirect (ID), implicit (IM) and implied compliments (ID and IM) taken together (IDM).

*Table 1: The number of compliments across languages and compliment types*

| | Total | D | IDM | ID | IM |
|---|---|---|---|---|---|
| English | 56 | 31 (c.55%) | 26 (c.46%) | 21 (c.38%) | 5 (c.9%) |
| Polish | 50 | 27 (54%) | 24 (c.48%) | 19 (38%) | 5 (10%) |

For a paper on subtitling compliments, *viz.* one ingrained in pragmatics, it might be perceived as a major omission if we turned a blind eye on the issue of the grammatical structure of compliments, even though this approach is definitely on the wane. Briefly, both the original soundtrack and the Polish translation have been analysed in terms of the linguistic structures used: fifty seven types of structures have been identified in the English text and fifty one in the Polish translation. Most structures were used once or twice only, within those employed more often the following have the highest frequency rates ('P' stands for 'pronoun'):

English: P+'be'+Adj (+P) (7); P+'look'+Adj (5); Adj+NP (3); P+V+NP (3);
Polish: P+'look'+Adj (5); Adj+NP (3); P+'be'+Adj (+P) (3) Adj (3); V+NP (3);

Compared with the structures claimed by Holmes (1986) to occur with highest frequency in New Zealand English, our data pertaining to American English (original soundtrack) are in agreement with what Holmes contended as far as the following structures are concerned: P+'be'+Adj and P+'look'+Adj. These structures appeared most often in the original soundtrack; however, their occurrences are not overwhelming (seven and five occurrences respectively). The Polish frequency list in turn is partially compatible with Manes and Wolfson's American English frequency list with P+'look'+Adj enjoying highest rate (five occurrences). These observations are interesting yet they are far from conclusive and generalisations cannot be proposed.

Now, in accordance with the more recent discursive approach to politeness[2], it is advisable that snippets of conversations be substituted by longer stretches

2 Alongside the general classification of compliments in line with the now classic theories, a more recent solution to discourse analysis known as the discursive approach is advocated by a growing number of scholars, promoted in particular by members of the British Linguistic Politeness Research Group (LPRG), and now spawning worldwide. Without going into details concerning the main theoretical-methodological foundations of this promising neo-Gricean (cf. Grice, 1975) and post-Brown and Levinsonian

of speech in discourse analysis. Indeed, in our material there are cases wherein labelling utterances with linguistic structures would not, beyond doubt, do full justice to the study. A linguistic analysis of praising stands in need of a more comprehensive approach; quintessentially, a compliment (in particular indirect or implicit) may span more than one sentence, and a straightforward expression of praise is not always the case, hence occasional difficulties in the identification of the key grammatical structure, as illustrated by the extract below:

(1)
English original soundtrack:
Scene 14 [subtitles no 281–283; onset time: 00:16:25,609]
281 At B.B.D. & O., Darcy led a creative team…
282 that snagged $500 million in new business wins.
283 And that was just last year alone.

Moreover, if we follow the paradigm of cognitive linguistics, finding the best fit for a compliment should not predicate on equivalence at lexical, phrasal or even at speech act level; rather the conceptual level of meaning should be the starting point for any comparisons and inferences (see Temmerman 2000 for the application of CL to terminology in specialised translation, and Tabakowska 2001 for the application of CL to literary interpretations). Inasmuch as the method of analysis inheres in assumptions voiced by cognitive linguistics, onomasiological, conceptual-driven equivalents a translator proposes (which necessarily also entail socio-pragmatic adjustments) may well depart from the original words and their morpho-syntactic properties. In view of this, following the template of grammatical structures offered by Manes and Wolfson (1980) or Holmes (1986) has low plausibility in analysing the translation of intercultural and interlingual compliments, because the lexical-grammatical (semasiological) facets constitute the focus of attention at the expense of the conceptual (onomasiological) facets.

Taken together, even though I was fully aware of the structure-oriented method of analysis being outdated, I was tempted to tabulate these differences to complete the study with yet another perspective only to eventually learn very little about the typological conventions and subtitler-specific preferences. Generally speaking, the analysis above shows that the number of compliments is lower in the Polish

(cf. Brown & Levinson, 1987) proposal (see LPRG (2011) for a full view of the paradigm and Bączkowska (2013) for its succinct overview in Polish), suffice it to say that it furnishes proof for above speech act analysis as the preferable way of discourse analysis, and in this respect it resembles conversation analysis methodology.

translation (which to some extent results from space and time constraints imposed on subtitles; see Díaz Cintas and Remael 2007: 84, 96) and that the grammatical structures used in both languages only partially overlap.

### *3.2.2 A cognitive linguistic approach to subtitles – construal operations*

In most general terms, cognitive linguistics challenges the traditional paradigm of truth-conditional semantics and allows situations (scenes) to be framed in a variety of ways, delegating the interpretation (construal) of a concept to the speaker. The concept of construal operations (Talmy, 1983 Langacker, 1987, 2008, Croft and Cruse, 2004) is defined by Langacker (2008: 43) as "our manifest ability to conceive and portray the same situation in alternate ways". For a subtitler, this approach to meaning opens up a range of possible renditions of a culture- and language-specific filmic discourse. From the inventory of construal operations distinguished by Talmy and Langacker, and their re-classified versions explored by Croft and Cruse, we shall focus on three of them applicable to the film translations under investigation, namely: perspective (vantage point), granularity and prominence.

#### *3.2.2.1 Perspective (vantage point)*

Perspective is a general term understood as the way a scene is viewed. In other words, it is a viewing arrangement, or, more specifically, the relationship between the viewer and what is viewed. Vantage point is a special case of viewing arrangement wherein scene interpretation depends on the location of the conceptualiser relative to the entity being viewed. For example, whether, say, a car is located in front of or behind a tree depends on the position of the speaker. Likewise, whether one says "come upstairs" or "go upstairs" depends on the location of the speaker (who may be upstairs or downstairs). Put differently, a perspectival construal imposes a specific interpretation of spatial relations. This construal operation is illustrated by sample 1.

1. [00:16:13,972 – 00:16:17,184]: Hey, standing room only. It's a first. (implicit compliment) // Brak miejsc siedzących [No seats left; lit. Lack of seats]; (implicit compliment).

Both the original text and the translated version adhere to the same compliment type. Considering this parameter only, there is no change between the English script and the rendered text. By resorting to construal operations it can be noticed that the basis for choosing a "standing perspective" in the English version is the speaker-oriented position (the speaker, i.e. the director of the company, Dan, is walking into a room full of employees sitting at a long table; he is just about to introduce a newly employed creative director, Darcy, to the employees). The Polish version, conversely, accommodates the "sitting position", i.e. the perspective stemming from the location of the employees gathered in the conference room. Viewing arrangement is thus reframed in the Polish rendition. In sum, the scene exemplifies a case wherein the typology of compliments alone does not do full justice to the difference between the original and the translated versions.

### *3.2.2.2 Granularity*

Granularity modifications are responsible for construing a scene at a fine-grained or coarse-grained level (Croft and Cruse, 2004: 51–53). For example, by saying "run through a field" one adopts a close (three-dimensional) view of the field, which allows one to see its thickness and internal details (vegetation), whereas in the case of "run across the field" a more surface-oriented (two-dimensional) picture is envisaged (Croft and Cruse, 2004: 51–53). By so doing, scalar adjustments are applied, which shift the level of specificity of the objects described.

In the example below, where Nick pays a compliment to his ex-wife, the construal is different in the subtitle than in the original text and it results in boosting granularity.

2. [00:13:39,805 – 00:13:43,767]: Honey, you look likc…$48 million. (direct compliment) // Wyglądasz jak million dolarów [You look like a million dollars]; (direct compliment).

"You look like a million dollars" is a conventional way of saying that someone looks exceptionally good, and the number as such does not make reference to any specific facet of the addressee. In the original soundtrack, however, the number is changed into 48 million dollars, and most probably it makes reference to the woman's age. With the use of a specific number, the appraisal seems more personal and precise (fine-grained); as a result, a higher resolution of the picture of a beautiful woman is achieved and thus admiration is reinforced. By the same token, focus is placed on the woman's unusually exceptional appearance. Owing to this shift in specificity, a marked version of the compliment is created. In the Polish version, by using the more recognisable "look like a million dollars", the compliment is more conventional, less marked and, consequently, attenuated. It is also more general (coarse-grained) as the number "one million" stands for some general quality of exceptional appearance.

In the next example, Darcy's (Nick's boss) mental faculty is described in highly positive terms in the English original text.

3. [01:54:12,590 – 01:54:15,844]: You are so much smarter than me. (direct compliment)// Jesteś mądrzejsza ode mnie [You are smarter than me]; (direct compliment).

The use of the intensifier *so much*, which modifies the graded adjective *smarter*, foregrounds the woman's exceptional intelligence. The woman is not only compared to the man but also her intelligence is clearly highlighted and specified. In the Polish version, on the other hand, *so much* is left out and, as a result, the subtitled version creates a sheer juxtaposition of the woman with the man, which makes the compliment less specific and less intense. The granularity of the compliment is thus smaller in the translated version than in the original soundtrack. These subtle differences in construal do not flow from the type of compliment used in English and Polish as both versions resort to direct compliment.

The last example in this section illustrates granularity achieved through the specification of *sir*, i.e. the scope of meaning of *sir* is narrowed to *sire* (lord) in the English version.

4. [00:09:32,704 – 00:09:34,914]: Good afternoon, sire. (direct compliment)// Dzień dobry panu [Good morning mister/sir] (direct compliment).

*Sire* used to be a form employed to address kings, so it is an emphatic, archaic and thus a marked version of *sir*. As originally it was used to show great respect towards the interlocutor of high rank, it is meant as a special compliment in this scene, even though it is used (by Nick's secretaries) with a vein of humour. The Polish subtitle, on the other hand, does not convey this shift in specificity and renders *sire*, typical of limited contexts, as *pan* (*mister/sir*) – a form which is commonly used in Polish and has a broader scope of meaning than an equivalent of *sire*. No special feeling of respect comparable to the meaning captured by *sire* is encoded by *pan*, hence the rendition is unmarked. The compliment is still sensed, however, as a weak indication of respect is conveyed by the formal term of address *pan* as opposed to the informal form normally used by the secretaries calling Nick by name. By resorting to a broad term *pan*, the resolution of the meaning is lower than in the English version, which, undoubtedly, leads to the loss of specificity (granularity) and, eventually, to compliment attenuation.

### *3.2.2.3 Prominence*

Prominence relates to our ability to attend to selected facets of a conceptualised entity. Put differently, it builds on a conceptual process which consists in focusing attention on a selected facet of a conceptualisation while ignoring other facets.

Giving prominence thus boils down to profiling conceptual content which is salient.

In the example below, pragmatically, the change between the two languages may be subsumed by the shift from indirect compliment in the original version to direct compliment in Polish.

5. [00:09:51,557 – 00:09:53,476]: Like a creative director.// Wymarzony dyrektor kreatywny.
   [A perfect creative director.]; (indirect > direct compliment).

The structure of the original utterance necessitates a reading of an indirect compliment due to the use of "like", which triggers a comparison between Nick and an imaginary director. The comparison is lost in the Polish translation inasmuch as "like" has been replaced by a straightforward label (you are) "a perfect director", which is suggestive of a fact rather than a comparison based on a non-fact encoded by (you look) "Like a perfect director". By highlighting the (assumed) present situation (i.e. being the creative director) rather than a possible future situation (being only a candidate for the creative director), a potential state is foregrounded in the English text, while the final state is profiled in the Polish subtitle (Nick is already envisaged as the creative director).

In the next example, the compliment "Good idea" is rendered as "Niezły pomysł" ("Not a bad idea").

6. [00:56:45,856 – 00:56:48,484]: Black and white could really help. Good idea, Nick. (direct compliment)// Czarno-białe będą lepsze. Niezły pomysł, Nick.
   [Black and white will be better. Not a bad idea, Nick.]; (direct compliment).

The English version carries more positive value expressed by the word "good", while in the Polish version the word "niezły" contains a negation of the adjective "bad" (the prefix *nie-* means 'not'). The subtitle thus profiles the utterance in a distinct way, with the subtitled compliment being less straightforward and weaker than in the original version, as "not bad" invokes, in the first place, the state of being "bad", and thus it is closer to the negative extreme on a cline of evaluation.

## 4. Conclusion

It has been the primary objective of this paper to study compliments in subtitles and, in particular, to shed some light onto a relatively novel approach to analyzing subtitles, namely through the lens of cognitive linguistics. As has been demonstrated, the juxtaposition of compliment types, the examination of their grammatical structure and the tabulation of the overall number of compliments occurring in the data under inspection have not proved sufficient for a precise delineation of the differences between the Polish renditions of compliments and their English original versions. Notably, the observations concerning compliment structures in Polish and English texts are inconclusive, while the number of compliments in the subtitles is only slightly smaller than in the original text (mostly resulting from the application of condensation, which is a recommended subtitling strategy). Therefore, in order to be more comprehensive and insightful, with a view to giving full justice to the way compliments are rendered via subtitles, it may be interesting and promising to employ a methodological framework advocated by cognitive linguistics. Construal operations appear to be particularly predisposed for this task and capable of offering a fine-grained approach to the analysis of compliments in interlingual subtitles. Arguably, the application of construal operations allows us to unearth additional information about how a compliment is conceptualised, which invites a more perspicacious examination of language; thus construal operations may complement the already existing methods of investigation.

## References


*Alvarez-Pereyre, M.* (2011). Using film as linguistic specimen: Theoretical and practical issues. In R. Piazza, M. Bednarek & F. Rossi (Eds.), *Telecinematic Discourse* (pp. 48–67). Amsterdam: John Benjamins.

*Bączkowska, A.* (2011). Some remarks on a multimodal approach to subtitles. *Linguistics Applied*, 4, 47–65.

*Bączkowska, A.* (2012a). Language and power in multimodal audiovisual translation. In P. Pęzik (Ed.), *Corpus Data across Languages and Disciplines* (pp. 267–280). Frankfurt am Mein: Peter Lang.

*Bączkowska, A.* (2012b). Multimodal analysis of im/politeness in subtitles. In M. Koszko, K. Kowalewska, J. Puppel & E. Wąsikiewicz-Firlej (Eds.), *Lingua: Nervus Rerum Humanarum. Essays in Honour of Professor Stanisław Puppel on the Occasion of his 65th Birthday* (pp. 61–78). Poznań: Adam Mickiewicz University Press.

*Bączkowska, A.* (2013). Podejścia dyskursywne w teorii grzeczności. In Pająkowska-Kęsik, M., Paluszak-Bronka, A. & Kołatka, K (Eds.), *Język, wielokulturowość, tożsamość* (pp. 19–33). Bydgoszcz: Bydgoskie Towarzystwo Naukowe.

*Bączkowska, A. & Kieś, M.* (2012). Multimodal subtitling of direct compliments in Polish, Italian and Swedish. In K. Kosecki & J. Badio (Eds.), *Cognitive Processes in Language* (pp. 41–51). Frankfurt am Mein: Peter Lang.

*Bonsigniori, V. Bruti, S. & Masi. A.* (2011). Formulae across languages: English greetings, leave-takings and good wishes in dubbed Italian. In A. Serban, A.

Matamala & J-M. Lavaur (Eds.), *Audiovisual Translation in Close-Up* (pp. 23–44). Frankfurt am Mein: Peter Lang.

*Boyle, R.* (2000). You've worked with Elisabeth Taylor! Phatic functions and implicit compliments. *Applied Linguistics*, 21(1), 26–46.

*Brown, P. & Levinson, S.* (1987). *Politeness: Some Universals in Language Usage*. Cambridge: CUP.

*Bruti, S.* (2006). Cross-cultural pragmatics: the translation of implicit compliments in subtitles. *The Journal of Specialised Translation*, 6, 184–197.

*Bruti, S.* (2009). The translation of compliments in subtitles. In J. Díaz Cintas (Ed.), *New Trends in Translation Studies* (pp. 226–238). Bristol: Multilingual Matters.

*Bruti, S. & Perego, E.* (2008). Vocatives in subtitles: a survey across genres. In C. Taylor (Ed.), *Ecolingua: The Role of E-corpora in Translation and Language Learning* (pp. 11–51). Trieste: Edizioni Università di Trieste.

*Caimi, A.* (2012). Cognitive insights into the role of subtitling in L2 learning. In A. Serban, A. Matamala & J-M. Lavaur (Eds.), *Audiovisual Translation in Close-Up* (pp. 113–129). Frankfurt am Mein: Peter Lang.

*Croft W. & Cruse, A. D.* (2004). *Cognitive Linguistics*. Cambridge: CUP.

*Deckert, M.* (2013). *Meaning in Subtitling. Toward a Constructive Cognitive Semantic Model*. Frankfurt am Main: Peter Lang.

*Delabastita, D.* (1989). Translation and mass-communication: film and TV translation as evidence of cultural dynamics. *Babel*, 35/4, 193–218.

*Díaz Cintas, J. & Remael, A.* (2007). *Audiovisual Translation: Subtitling*. Manchester: St. Jerome.

*Freddi, M.* (2008). The phraseology of contemporary filmic speech: formulaic language and translation. In M. Freddi & M. Pavesi (Eds.) *Analysing Audiovisual Dialogue: Linguistic and Translational Insights* (pp. 101–124). Bologna: CLUEB.

*Gambier, Y.* (2006). Multimodality and audiovisual translation. *MuTra2006–Audiovisual translation scenarios*. http://www.euroconferences.info/proceedings/2006_Proceedings/2006_Gambier_Yves.pdf.

*Goffman, E.* (1967). *Interaction Ritual – Essays in Face to Face Behaviour*. New York: Aldine Publishing.

*Gottlieb, H.* (2008). Multidimensional translation. In A. Schjoldager, H. Gottlieb & I. Klitgård (Eds). *Understanding Translation* (pp. 39–65). Aarhus: Academica.

*Grice, P.* (1975). Logic and Conversation. In P. Cole & J. Morgan (Eds.) *Syntax and Semantics* (pp. 41–58). New York: Academic Press.

*Hatim, B. & Mason, I.* (1996). *The Translator as Communicator*. London: Routledge.

*Herbert, R.* (1991). The sociology of compliment work: an ethnocontrastive study of Polish and English. *Multilingua*, 10(4), 381–402.

*Holmes, J.* (1986). Compliment and compliment responses in New Zealand English. *Anthropological Linguistics*, 28(4), 485–508.

*Jaworski, A.* (1995). This is not an empty compliment: Polish compliments and the expression of solidarity. *International Journal of Applied Linguistics*, 5(1), 63–94.

*Jucker, A. H.* (2009). Speech act research between armchair, field and laboratory. The case of compliments. *Journal of Pragmatics*, 41, 1611–1635.

*Kerbrat-Orecchioni, C.* (1987). La description des échanges en analyse conversationelle: l'example du compliment. *DRLAV- Revue de Linguistique*, 36(37), 1–53.

*Langacker, R.* (1987). *Foundations of Cognitive Grammar*. Vol. 1. *Theoretical Prerequisites*. Stanford: Stanford University Press.

*Langacker, R.* (2008). *Cognitive Grammar: A Basic Introduction*. Oxford: OUP.

*Lewandowska-Tomaszczyk, B.* (1989). Praising and complimenting. In W. Oleksy (Ed.) *Contrastive Pragmatics* (pp. 73–100). Amsterdam: John Benjamins.

*Linguistic Politeness Research Group (LPRG).* (2011). *Discursive Approaches to Politeness*. Berlin: Mouton de Gruyter.

*Manes, J. & Wolfson, N.* (1980). The compliment formula. In F. Coulmas (Ed.) *Conversational Routine* (pp. 115–132). The Hague: Mouton.

*Matamala, A & Lorente, M.* (2008). New tools for translators: INTCA, an electronic dictionary of interjections. In D. Chiaro, C. House & C. Bucaria (Eds.) *Between Text and Image: Updating Research in Screen Translation* (pp. 63–75). Amsterdam: John Benjamins.

*Pavesi, M.* (2008). Spoken language in film dubbing: target language norms, interfaces and translational routines. In D. Chiaro, C. Heiss & C. Bucaria (Eds.) *Between Text and Image: Updating Research in Screen Translation* (pp. 79–100). Amsterdam: John Benjamins.

*Pavesi, M.* (2009). Dubbing English into Italian: a closer look at the translation of spoken language. In J. Díaz Cintas (Ed.). *New Trends in Audiovisual Translation* (pp. 195–209). Bristol: Multilingual Matters.

*Piazza, R. Bednarek, M., Rossi, F. (Eds).* (2011). *Telecinematic Discourse. Approaches to the language of films and television series*. Amsterdam: John Benjamins.

*Rosa, A.* (2001). Features of oral and written communication in subtitling. In Y. Gambier & H. Gottlieb (Eds.) *(Multi) Media Translation* (pp. 213–222). Amsterdam: John Benjamins.

*Rose, K. R.* (2001). Compliments and compliment responses in film: implications for pragmatics research and language teaching. *International Review of Applied Linguistics. Language Teaching* 39/4, 309–326.

*Rossi, F.* (2011). Discourse analysis of film dialogues. Italian comedy between linguistic realism and pragmatic non-realism. In R. Piazza, M. Bednarek &

F. Rossi (Eds.). *Telecinematic Discourse. Approaches to the Language of Films and television Series* (pp. 21–46). Amsterdam: John Benjamins.

*Skuggevik, E.* (2009). Teaching screen translation: the role of pragmatics in subtitling. In J. Díaz Cintas and G. Anderman (Eds.) *Audiovisual Translation: Language Transfer on Screen* (pp. 197–213). Basingstoke: Palgrave Macmillan.

*Tabakowska, E.* (2001). *Językoznawstwo kognitywne a poetyka przekładu*. Kraków: Universitas.

*Talmy, L.* (1983). How language structures space. In H. L. Pick Jr. & L. P. Acredolo (Eds.) *Spatial Orientation: Theory, Research and Orientation* (pp. 225–282). New York: Plenum Press.

*Taylor, C.* (2003). Multimodal transcription in the analysis, translation and subtitling of Italian films. *The Translator*, 9/2, 191–205.

*Taylor, C.* (2004). Multimodal text analysis and subtitling. In E. Ventola, C. Charles & M. Kaltenbacher (Eds.). *Perspectives on Multimodality* (pp. 153–172). Amsterdam: John Benjamins.

*Taylor, C.* (2009). Pedagogical tools for the training of subtitlers. In G. Anderman & J. Díaz Cintas (Eds.). *In So Many Words: Dubbing and Subtitling for the Screen* (pp. 214–228). Clevedon: Multilingual Matters.

*Temmerman, R.* (2000). *Towards New Ways of Terminology Description*. Amsterdam: John Benjamins.

*Tomaszkiewicz, T.* (2006). *Przekład audiowizualny*. Warsaw: PWN.

*Tomaszkiewicz, T.* (2009). Linguistic and semiotic approaches to audiovisual translation. In M. Freddi & M. Pavesi (Eds.) *Analysing Audiovisual Dialogue:Linguistic and Translational Insights* (pp. 19–29). Bologna: CLUEB.

*Wolfson, N.* (1984). Pretty is as pretty does. A speech act view of sex roles. *Applied Linguistics*, 5(3), 236–244.

*Wolfson, N. & Manes, J.* (1980). The compliment as a social strategy. *Papers in Linguistics: International Journal of Human Communication*, 13 (3), 391–410.

*Yuan, Y.* (2002). Compliments and compliment responses in Kunming Chinese. *Pragmatics*, 12/2, 183–226.

*Yuan, X.* (2012). *Politeness and Audience Response in Chinese-English Subtitling*. Frankfurt am Mein: Peter Lang.

Natalya Savoiskaya
*University of Santiago de Compostela, Spain*

# Specific universals: a comparative analysis of subject of evaluation construal

Financial support for this article has been provided by Erasmus Mundus Action II Marco XXI Project, grant No 20112572.

**Abstract:** This study is aimed at discovering similarity and specificity in evaluation construal, with a special emphasis on the subject of evaluation. As viewed by Wolf (2002), the subject can voice individual appraisal or refer to common ground. To differentiate between the types of subject construal and to allow for the comparative quantitative analysis in English and Russian, the notions of individual, mutual, and social evaluation are introduced. Considering the explicitness and implicitness of the units analysed, the method of random sampling from fiction texts was preferred over the other methods, as it allows for a deeper conceptual analysis of implicative structures, and, as is demonstrated in the selected examples, hidden components of meaning are often characteristic of utterances of appraisal.

Keywords: subject of evaluation, explicit/implicit evaluation, individual/mutual/social evaluation, English/Russian.

## 1. Linguistic universalism

In publications on universalism in language, we find different opinions on this particular issue. First, generativists consider the grammatical diversity of languages as "largely superficial," whereas the analysis of the abstract language levels demonstrates a similarity, to a great degree (Langacker, 1987, p. 46). As Palmer (1996) summarises it, there is plenty of evidence both for and against universalism in language; both cognitive linguists and anthropologists, studying the mental processes of conceptualising, categorising, and schematising have found enough examples of how languages can function in similar ways (p. 117).

The question of universality has been raised in numerous works: Wierzbicka (1999), Wolf (2002), Kornylov (2003), Apresyan (1995), Kolshansky (1990). Wierzbicka (1999) emphasises the existence of fundamental notions lexicalised in every language, which constitute the conceptual basis for human language and cognition, and exist side by side with a great number of those that are culturally specific (pp. 321–322). Kornylov (2003) introduces the term, "psychological unity of the humankind," for the conceptual basis of human consciousness essential for cross-cultural understanding (p. 140). Apresyan (1995) and Potebnya (1976) acknowledge the existence of the universal conceptual basis but specify that people construe reality "through the prism of their languages" (p. 39; p. 260).

The underlying idea of this research is that the human brain is innately universal, but, when placed in different surroundings and with the perception of different world views, the universal categories are varied to some degree. An attempt is undertaken to exemplify the coexistence of the universal and specific content on the language of evaluation. Evaluation is conceptualised by people in any culture; being evaluated can mean being an object of hunting by a tribesman in an African bush as well as of the consequences of the nuclear disaster in Fukushima; or, a poet can evaluate the beauty of a fair lady, and we all evaluate the taste of honey. The specificity is in the linguistic representation of appraisal.

## 2. Evaluation in language: approaches to understanding

The very idea of evaluative language studies is certainly not new as there have been works on linguistic evaluation that uncover its different aspects—among others, the structural organisation of evaluative language (Wolf, 2002; Martin & White 2005), the specificity of evaluative meanings construal in different languages (Taboada, Carretero, & Hinel, 2014, pp. 127–161), as well as in different registers of one language (Biber, 2006, pp. 97–116; Carter, 2006). The idea of automatic identification of appraisal units for the analysis of a wide corpus of evaluative constructions has also been put forward (Taboada et al., 2014, pp. 127–161; Hunston, 2011). Nonetheless, evaluation, being a very complicated issue to study, continuously draws researchers' attention, offering many directions for investigation.

It should be mentioned that the very notion of evaluation in linguistics is perceived differently by different scholars. As Hunston (2011) notes, "a wide brand of terminology is used to refer to similar areas of language use: appraisal, stance, sentiment, evaluative, attitudinal or affective language, metadiscourse and evaluation" (p. 10); all of them can be treated as synonymous, since the differences are very peculiar to a certain type of study.

There are two primary theoretical frameworks that inform this study with regard to the language of evaluation. The first is *"Functional Semantics of Evaluation"* by Wolf (2002), who draws attention to the structure of evaluation and its elements, with her aim being "to give an idea of the place occupied by evaluative meanings in the language units and speech structures" (p. 5). The second is *"The Language of Evaluation,"* by Martin and White (2005), who work within systemic functional linguistics (SFL) and offer the engagement/attitude/graduation framework. The basic idea underlying both frameworks is that the main components of evaluation units are the evaluator and the linguistic means of evaluation, which are identified in any language.

## 2.1 Subject of evaluation in Functional Semantics and Systematic Functional Linguistics

In this study special attention is given to the subject of evaluation. Wolf (2002) considers the subject and object to be the main elements of an evaluative item, connected by an evaluative predicate. The subject of evaluation is a person or society from whose point of view something is evaluated, whereas the object is what is evaluated (p. 12). The evaluative predicate is the main means of expressing an evaluation; it connects its subject and object (p. 97). Wolf points out several ways of indicating (or not indicating) a subject; among them, a) explicit vs. implicit ones, and b) individual, referring to the conceptual world of the evaluator, vs. social ones, presuming the enclosure of opinions comprising social common ground.

In Martin and White's work (2005) the component of meaning that refers to the voice of the evaluative structure is known as engagement that "deals with sourcing attitudes and the play of voices around opinions in discourse" (p. 35). In the framework of SFL, distinction is made between monoglossic and heteroglossic engagement, and the criteria for it is the dialogistic perspective, i.e., if an utterance makes reference to outer voices or opinions, it is heteroglossic; otherwise it is monoglossic (pp. 99–100).

Though the idea of the voice of evaluation is basically similar, the angle from which an evaluator is viewed introduces certain differences. Thus, speaking about individual evaluation, Wolf (2002) emphasises the affectivity and partiality of individual evaluation on the one hand, and on the other, describes phrases like *in my opinion, in his opinion* as a means of explicating the subject, because it refers the evaluative utterance to the person's conceptual world (p. 70). Thus, for example, the following sentence could be classified as a sample of individual evaluation: "*In my view* [italics added] the banks have been greedy" (Martin & White, 2005, p. 100). In Martin and White's terms, it is an example of the heteroglossic, as the words, "in my view," presume the existence of other views, and in this way the speaker proposes a further dialogue.

In this study, judgments are based on Wolf's proposal of individual and social evaluation, as appraisal is viewed from the point of view of its linguistic construal and the reference to common ground knowledge.

### 2.2 Explicit vs. implicit evaluation

Dwelling on complexity of language of appraisal, Hunston (2011) emphasises that "every text and every utterance is evaluative" (p. 19). This idea is well grounded by the correlation of linguistic and mental content of the utterance, with the mental elements present only in the conceptual world—in the mind of the speaker. Hence, there are elements in the utterance, i.e., explicit, and others, hidden behind words and decoded at the level of consciousness, provided that the speaker and interlocutor share common conceptual worlds; in the case where they do not, an utterance can lead to a misunderstanding. This type of evaluative construction is known as either implicit evaluation (Hunston, 2011; Wolf, 2002; Markelova, 1993) or evoked evaluation (Martin, 2000).

In explicit constructions, evaluation can be easily identified by the presence of appraisal lexis, whereas the implicit ones can be devoid of such vocabulary, and the evaluative meaning is uncovered either through the context or due to the community of conceptual worlds, e.g., "*You want tickets the rest of your life?* [italics added] Your inspection ticket's expired. You realize that? One of your tail-lights is burned out. You aware of that?" (Cornwell, 2007, p. 33). It is, on the one hand, the preceding and following utterances that make the reader understand the situation, infer what kind of tickets are implied, and, consequently, evaluate the utterance as negative rather than neutral; on the other hand, from the context, it is common conceptual ground shared by the policeman and the driver that makes them understand each other. Examples of how neutral words can be evaluative are also provided by Hunston (2011, pp. 112–116) and Wolf (2002, pp. 47–67).

Implicitness is often characteristic of the subject of evaluation. Any pronounced evaluative unit reflects the conceptual world of its author; every person belongs to a society and, consequently, his conceptual world is a combination of societal and individual features. As a result, when he evaluates something, he bases his judgment on the general knowledge and moral of his society. In some societies being talkative is welcome and appreciated as positive, and silent visitors are considered gloomy and indifferent, whereas in other societies they may be regarded as garrulous and even impolite.

## 3. Comparative analysis of subject of evaluation construal in Russian and English

### 3.1 Process and methods of analysis

The material for this research is drawn from works of fiction. The evaluation units for the analysis were selected by random sampling and then annotated manually.

It can be argued that works of fiction are a reflection of the author's world view and the characters are the result of the writer's imagination, and the whole message of the story is, therefore, authorial; consequently, it is the author's language of evaluation that is analysed. Nevertheless, every author is a bearer of his cultural common ground and a follower of its values and, therefore, they are necessarily reflected in his writings.

Both the necessity to give consideration to implicit evaluative units and the difficulties connected with the analysis of larger social contexts containing evaluative meanings, have limited to a degree, the random sampling of evaluative utterances. Moreover, the selection is determined by the aims of a future endeavour to study cultural implications in communicative behaviour; thus, it is restricted to dialogical speech. Moreover, it is limited by the aim of the present research to focus on the subject of evaluation. Hence, the language of evaluation in this study is represented by samples of minimal stretches of dialogical speech containing any language units with evaluative semantics or directly implying it. By minimal stretches of speech, I am referring to excerpts that make the evaluative meanings clear to the reader; they require neither any special knowledge, nor a wider cultural context. Consider the following examples:

(1) It's a great article. (Cleave, 2008, p. 51)
(2) Ах, какие вы прекрасные ребята! (Prilepin, 2010)
Ah, what nice guys you are! [1]

In these sentences the words with evaluative semantics, *great* and *прекрасные (nice)*, make the appraisal quite evident. Examples (3) and (4) are devoid of evaluative semantics, which, however, does not prevent us from recognising the speaker's evaluative intent.

(3) "Ah, yes! Paris in the twenties! How much I learned there, how much I learned. How much I ate!" (Barlow, 2004, p. 18)

The combination of the interjection, the exclamation marks, and the repetition of *how much* contribute to our understanding of the positive character of the utterance.

(4) Какая разница, за кого голосовать? (Filippov, 2012)
What difference does it make for whom to vote?

1 Translations into English are my own.

This statement has a negative, rather than neutral, connotation, considering, first, that indifference in itself is a negative attitude and, second, that the election towards which one is indifferent is hardly worth praise.

As this research is connected with quantitative analysis, it is necessary to give more detail on the procedure of random sampling. The exemplifying sentences, repeated by the same character in the same situation, were not taken into account as they do not reflect the frequency of use of the evaluative unit in the society, but rather characterise the person's emotional state. Furthermore, emotional imperatives are not included in the analysis, evaluative though they are (for example, *Shut up!*), on the ground that the subject of evaluation in them is always the speaker, and this fact supposedly introduces an imbalance and enlarges the group of samples with an individual subject.

To mark the subject of evaluation, the cognitive approach was used. With regard to the works in cognitive linguistics (Langacker, 1987; Palmer, 1996; Popova & Sternin, 2007), cognitive interpretation presumes mental generalisation at an abstract level to describe cognitive characteristics of language units; this research is aimed at distinguishing three voices—individual, social and mutual—in the subject of evaluation; the detailed explanations of the analysis are given in Chapter 3.2.

In total, 310 examples of evaluation units in Russian and 332 in English have been analysed; they were extracted from the works of fiction by contemporary authors writing in Russian and in English: Prilepin, Andreyev, Zakashanskiy, Osipov, Filippov, Barlow, Lodge, Cleave, Cornwell, and Gilbert.

## 3.2 Criteria of classification

Considering a subject as a "person, part of society or society as a whole from the point of view of which evaluation is presented," and taking into account that "the subject evaluates on the basis of evaluation scale and situational stereotypes in his world view" (Wolf, 2002, p. 68), this analysis emphasises the degree of common ground involvement in the evaluative utterance. As the study has shown, utterances when the subject of evaluation does not coincide with the speaker are not rare, e.g.:

(5) Funerals are always like that. All the old skeletons come theatrically out of their closets. (Cleave, 2008, p. 127)

We hear the "voice" of social opinion rather than the speaker's. On the contrary, in

(6) I mean exploiting *me*, not the customers. (Lodge, 1975, p. 33)

the speaker quite clearly states her own opinion. Moreover, in an act of evaluation there might be a situation when the subject voices his opinion, but, at the same time, involves interlocutors, presuming they have the same opinion or encouraging them to join him in his views, for example, in

(7) I had, very early on, developed a penchant for beluga caviar, which could, as you can imagine, have perilous financial implications. (Barlow, 2004, p. 16)

*as you can imagine* is the signal that the interlocutor is supposed to infer the same from the situation.

When analysing the subject in an evaluative structure it is important to understand whose conceptual world is represented and how. According to Wolf (2002), three kinds of opinions can be distinguished: the speaker's opinion, social opinion, and the addressee's opinion (p. 70). In the current analysis they are treated as "voices," and are called *individual*, or the speaker's voice, *mutual*, or the voice of the speaker and interlocutor together, and *social*, where the opinion of many is stated through the speaker's voice, the main criterion for classification is the degree of shared views involved in the evaluation utterance.

Though random sampling is time-consuming, it allows for the selection of implicit evaluation units. The examples were picked up and marked manually as *individual*, *mutual* or *social* evaluation.

### 3.2.1 Individual evaluation

Samples of individual evaluation are the most numerous in both English and Russian. The individual voice is stated explicitly, or it can be implied; in this case it is the researcher's responsibility to recognise its type. Explicit wording includes various means in different languages. In Russian, Wolf (2002) distinguishes phrases with direct reference to a person, for example, *по-моему (in my opinion), по его мнению (in his opinion)*, etc., and propositional verbs like *полагать (suppose), считать (think, consider),* etc.; similar means of person reference can be used in English, e.g.:

(8) И считаю это верным. (Prilepin, 2010)
And I believe it's right.

(9) Я постарел. (Prilepin, 2010)
I've got older.

(10) I'm not surprised your studies are suffering. (Lodge, 1975, p. 50)

(11) Jolly decent of him to think of me, I must say. (Lodge, 1975, p. 24)

The most numerous direct way of person reference in both Russian and English is personal and possessive pronouns, for example:

(12) Мне так неудобно… (Prilepin, 2010)
It is inconvenient for me…

(13) Любимая моя сказка… (Prilepin, 2010)
My favourite fairy tale…

(14) Listen, kid, let me give you some fatherly advice. (Lodge, 1975, p. 33)

(15) And that comes with my sincere gratitude, young man, to be sure it does. (Barlow, 2004, pp. 8–9)

An interesting and rather complicated issue is the case with direct questions addressing the interlocutor, in which we can hear two voices. First, it is the speaker's intention to evaluate; nonetheless, the question may be non-evaluative as such, but on the other hand, it is an explicit engagement to dialogue as it encourages the interlocutor to formulate his appraisal, e.g.:

(16) Тебе нравится Хэми? (Prilepin, 2010)
Do you like Hammy?

(17) Students getting you down? (Lodge, 1975, p. 43)

In negative questions, though, the voice of the speaker is much louder, and it prevails over his opponent's voice, e.g.:

(18) Тебя не огорчает, что тебе уже двадцать лет, а ты по-прежнему полный дебил? (Prilepin, 2010)
Don't you care that you are twenty already, but still a complete moron?

(19) Don't you think this is a great view? (Lodge, 1975, p. 80)

Nevertheless, the speaker may agree or disagree with this evaluation; in case he agrees, their voice becomes mutual; otherwise, it will be a separate act of evaluation with another subject.

Special questions are a different case, when the speaker formulating a question presumes that the interlocutor is of the same opinion, and only has to choose a variant to clarify the state of things, for example:

(20) Ты что такой похнюпый? (Prilepin, 2010)
Why are you so sad?

(21) Why are you so keen, anyway? (Lodge, 1975, p. 67)

As mentioned above, units of individual evaluation are the most numerous in both languages, but the difference is in the number of explicitly and implicitly stated individual opinions. In Russian only 14.3% of all units of individual evaluation are those where the subject is stated explicitly, whereas in English the percent is higher, 41.5%; 73.6% and 45.0%, respectively, are samples where the individual subject is implied; for example:

(22) Даже не город, поселок. (Osipov, 2012)
Not even a town, a village.

(23) You look beautiful today, *signorina*… (Gilbert, 2006, p. 70)

As such samples are taken from dialogues, it is inferred, though not stated directly, that the appraisal is the opinion of one of its participants.

Furthermore, among units of individually expressed evaluation, the following classes were identified:

- individual evaluation based on commonly accepted knowledge (10.6 % in Russian and 3.7% in English), e.g.:

(24) But that's ridiculous, *Playboy* isn't *pornography*, for heaven's sake! Why, clergymen read it. Clergymen write for it! (Lodge, 1975, p. 105)

(25) Это не по-человечески! (Zakashanskiy, 2012)
It's not humane!

- individual evaluation with the speaker distancing himself from the opinion pronounced. This affect is achieved by using the second person singular pronoun for generalising the idea—verbs like *seem*, *appear*, etc.—and subjunctive mood constructions, e.g.:

(26) It is easier when you are from outside. (Cleave, 2008, p. 64);

(27) All appears to be in order. Quite acceptable. (Barlow, 2004, p. 27)

(28) After all, for the better part of a decade a great many of the things I had sent down to my stomach could have been called food only through a very liberal understanding of the word, or by a desperately hungry person: in Oxford I had feasted on a sturdy hiking boot… (Barlow, 2004, p. 20)

Constructions of this type were found in the English research material only and constituted 5.8 %.

- Individual voice rendering the opinion of a third party (1.5% in Russian and 4% in English), e.g.:

(29) Он рассказывает о том, что ему не дают ставить спектакль. Что его обижают. Ему не с кем общаться… (12) (Prilepin, 2010)
He says that they do not allow him to stage the performance. That they offend him. He has nobody to communicate with.

(30) Bernadette being her niece, she feels responsible for the girl's moral welfare. (Lodge, 1975, p. 105)

Although individual evaluation represents the speaker's voice, the analysis has shown that it is far from being homogeneous. We can only speak of the purely individual voice in case the subject of evaluation is stated explicitly or if evaluation is affective, as social opinions are normally stated in a more reserved way. Otherwise, individual appraisal may reflect a social voice, encourage the interlocutor to share his views, or the speaker, when voicing his opinion, may distance himself from this view. As research has shown, in Russian and English there are similar ways to construe the individual subject of evaluation, but the frequency of their use is not the same.

### *3.2.2 Mutual evaluation*

In the previous section some samples of interlocutor's enclosure in the evaluative structure were given. In case of questions, both general and special, we observe the implication of the interlocutor's evaluative views. There is also an explicit way to represent the enclosure of the interlocutor's cognitive world in evaluation structures. Such examples were grouped into a separate class of mutual evaluation.

In the two languages, a typical construction of mutual evaluation is to address the interlocutor as if asking for his approval by using words like *помнить, слушать, заметить, imagine, know, believe,* etc.; for example:

(31) И, заметь, они никогда не хамили старшим. (Alexeyev, 2013)
And you know they have never been rude to the elders.

(32) It appears, gentlemen, that you find the act of eating a chair quite ordinary, quite … beneath your contempt indeed. (Barlow, 2004, p. 49)

Besides the similarity, there are different means of addressing the second person. For example, tag-questions are rather popular in English:

(33) They certainly don't mince their words, do they? (Lodge, 1975, p. 68)

(34) You don't even know what planet you're on, do you? (Cleave, 2008, p. 88)

Tag-questions express the speaker's certainty about the coincidence in the evaluation pattern; they sound more like statements rather than questions. In Russian, the very idea of a tag does exist, but it is rendered differently, with the help of words like *да, правда, а, что*:

(35) Здорово, да? (Prilepin, 2010)
Nice, isn't it?

(36) Что, неприятно? (Zakshanskiy, 2012)
Distressing, isn't it?

In addition, a few examples of a complex object were found—suppositional *must* and invitational *давай (let's)* that also reflect the idea of the second person's involvement in the evaluative situation; for example:

(37) "This man, gentleman, believes me to be a liar." (Barlow, 2004, p. 32)

(38) You must be all excited then. (Lodge, 1975, p. 39)

(39) Давай соврем что-нибудь? (Prilepin, 2010)
Let's tell lies?

### *3.2.3 Social evaluation*

In units of this class, the voice of society or a social group is too strong and overwhelms the voice of the speaker.

(40) Seriously? Out in public? With your *husband*? Isn't that terribly last season? (Cleave, 2008, p. 52)

(41) Это ужасно, что я хочу водки? (Prilepin, 2010)
Is it terrible that I want vodka?

Units of social evaluation are not numerous, but they should be distinguished among the others. Here, common knowledge, where the voices of the speaker and society intersect, is the basis for understanding. Very often, though, the question of precedence of voices can be open to debate. Thus in (41) the speaker's opinion is based on the commonly denounced action of drinking vodka. However, dwelling on the issue of evaluation construal, one can argue that it is not the very fact of drinking vodka that the person evaluates; rather, for her—a young woman—it seems terrible that she might have the desire to do so; in this case, it is the individual that prevails, though based on social opinion.

## 4. Conclusion

The study of subject of evaluation construal in Russian and English has shown that in both languages there are patterns that reflect individual and social voice. They can intersect, and at times, it is disputable whether the evaluation is purely individual or exclusively social. For the intersection of the speaker's and the interlocutor's voices, the term of mutual voice has been introduced; it is believed to reflect the idea of the second person's involvement in the evaluation process; with it, it is distinct from social evaluation in the sense that it is not necessarily based on common ground and can be quite effective. It is also distinct from Martin and White's (2005) heteroglossic engagement because it usually presumes support of the speaker's opinion, whereas heteroglossy is generally viewed as an invitation to dialogue.

The analysis has shown more similarities than differences in the ways of stating who the evaluator is. Nonetheless, certain differences were found in the ratio of the explicit and implicit voicing and linguistic means used. The significant difference found is that of the explicitly stated subject; in English it is much more frequent when the speaker pronounces himself to be the evaluator with the help of pronouns; in Russian such examples are fewer in number and the grammar can be construed differently; with the synthetic nature of the Russian language permitting, personal pronouns are not necessary. The minor differences in the construal of evaluation units can be explained by different morphology of the languages studied.

To sum up, the question of the proportion of the individual to common in the subject of evaluation is open to discussion; this research has attempted to distinguish between different types of subject of evaluation construal to enable a comparative study. However, it is acknowledged that there might be other criteria for classification, e.g., a dialogistic perspective as offered by Martin and White (2005), or a combination of the criterion of heteroglossy and that of common ground involvement.

## References

*Apresyan, Y. D.* (1995). Образ человека по данным языка: попытка системного описания [Image of a person by language data: An attempt of systematic description]. *Вопросы языкознания* [Issues in Linguistics], *1*, 37–67.

*Biber, D.* (2006). Stance in spoken and written university registers. *Journal of English for Academic Purposes, 5*, 97–116.

*Carter, R., & McCarthy, M.* (2006). *Cambridge grammar of English*. Cambridge: Cambridge University Press.

*Hunston, S.* (2011). *Corpus approaches to evaluation. Phraseology and evaluative language*. New York, London: Routledge.

*Kolshansky, G. V.* (1990). *Объективная картина мира в познании и языке.* [Objective world picture in cognition and language]. Moscow: Nauka.

*Kornylov, O. A.* (2003). *Языковые картины мира как производные национальных менталитетов.* [Language world pictures as derivatives of national mentalities]. (2nd ed.). Moscow: CHeRo.

*Langacker, R.* (1987). *Foundations of cognitive grammar. Volume I. Theoretical prerequisites.* Stanford: Stanford University Press.

*Markelova, T. V.* (1993). *Семантика оценки и средства ее выражения в русском языке.* [Semantics of evaluation and means of its expression in Russian]. Moscow: MPU.

*Martin, J. R.* (2000). Beyond exchange: APPRAISAL systems in English. In S. Hunston & G. Thompson (Eds.), *Evaluation in texts: Authorial stance and the construction of discourse* (pp. 142–175). Oxford: Oxford University Press.

*Martin, J. R., & White, P. R. R.* (2005). *The language of evaluation.* New York: Palgrave.

*Palmer, G. B.* (1996). *Toward a theory of cultural linguistics.* Austin: University of Texas Press.

*Popova, Z. D., & Sternin, I. A.* (2010). *Когнитивная лингвистика.* [Cognitive linguistics]. Moscow: AST: Vostok – Zapad.

*Potebnya, A. A.* (1976). *Эстетика и поэтика.* [Aesthetics and poetics]. Moscow: Iskusstvo.

*Taboada, M., Carretero M., & Hinnell J.* (2014) Loving and hating the movies in English, German and Spanish. *Languages in Contrast, 14* (1), 127–161.

*Wierzbicka, A.* (1999). *Семантические универсалии и описание языков. [Semantic universals and description of languages].* (A. D. Shmelyev, Trans.). Moscow: Yazyki russkoy cultury.

*Wolf, E. M.* (2002). *Функциональная семантика оценки.* [Functional semantics of evaluation]. Moscow: URSS.

## Works of fiction

*Alexeyev, I.* (2013). *Костюм.* [The Costume]. Retrieved from http://magazines.russ.ru/znamia/2013/1/a4.html

*Barlow, J.* (2004). *Eating Mammals.* New York: HarperCollins *Publishers.*

*Cleave, C.* (2008). *The other hand.* UK: Sceptre.

*Cornwell, P.* (2007). *At risk.* UK: Sphere.

*Filippov, D.* (2012). *Стратегия 19.* [Strategy 19]. Retrieved from http://magazines.russ.ru/znamia/2012/5/f8.html

*Gilbert, E.* (2010). *Eat. Pray. Love.* London: Bloomsbury.

*Lodge, D.* (1975). *Changing Places.* Penguin Books.

*Osipov, M.* (2012). *Домашний кинотеатр.* [Home theatre]. Retrieved from http://magazines.russ.ru/znamia/2012/5/o4.html

*Prilepin, Z.* (2010). *Грех и другие рассказы.* [The Sin and other stories]. Retrieved from http://loveread.ws/view_global.php?id=11947

*Zakashanskiy, K.* (2012). *Бомба.* [The Bomb]. Retrieved from http://magazines.russ.ru/znamia/2012/5/z6.html

**ŁÓDŹ**
**Studies in Language**

Edited by
Barbara Lewandowska-Tomaszczyk and Łukasz Bogucki



Vol. 1 Barbara Lewandowska-Tomaszczyk / Patrick James Melia (eds.): PALC'99: Practical Applications in Language Corpora. Papers from the International Conference at the Universtiy of Łódź, 15-18 April 1999. 2000.

Vol. 2 Lou Burnard / Tony McEnery (eds.): Rethinking Language Pedagogy from a Corpus Perspective. Papers from the Third International Conference on Teaching and Language Corpora. 2000.

Vol. 3 Barbara Lewandowska-Tomaszczyk / John Osborne / Frits Schulte: Foreign Language Teaching and Information and Communication Technology. 2001.

Vol. 4 Piotr Cap: Explorations in Political Discourse. Methodological and Critical Perspectives. 2002.

Vol. 5 Ewa Waniek-Klimczak / Patrick James Melia (eds.): Accents and Speech in Teaching English Phonetics and Phonology. EFL perspective. 2002.

Vol. 6 Barbara Lewandowska-Tomaszczyk / Kamila Turewicz (eds.): Cognitive Linguistics Today. 2002.

Vol. 7 Barbara Lewandowska-Tomaszczyk (ed.): PALC 2001: Practical Applications in Language Corpora. 2003.

Vol. 8 Corpus Linguistics by the Lune. A Festschrift for Geoffrey Leech. Edited by Andrew Wilson, Paul Rayson and Tony McEnery. 2003.

Vol. 9 Barbara Lewandowska-Tomaszczyk (ed.): Practical Applications in Language and Computers. PALC 2003. 2004.

Vol. 10 Imagery in Language. Festschrift in Honour of Professor Ronald W. Langacker. Edited by Barbara Lewandowska-Tomaszczyk and Alina Kwiatkowska. 2004.

Vol. 11 Seana Coulson / Barbara Lewandowska-Tomaszczyk (eds.): The Literal and Nonliteral in Language and Thought. 2005.

Vol. 12 Piotr Cap (ed.): Pragmatics Today. 2005.

Vol. 13 Jacek Waliński / Krzysztof Kredens / Stanisław Goźdź-Roszkowski (eds.): Corpora and ICT in Language Studies. PALC 2005. 2007.

Vol. 14 Krzysztof Kosecki (ed.): Perspectives on Metonymy. Proceedings of the International Conference "Perspectives on Metonymy", held in Łódź, Poland, May 6-7, 2005. 2007.

Vol. 15 Barbara Lewandowska-Tomaszczyk / Tomasz Płudowski / Dolores Valencia Tanno (eds.): The Media and International Communication. 2007.

Vol. 16 Krzysztof Kredens / Stanisław Goźdź-Roszkowski (eds.): Language and the Law: International Outlooks. 2007.

Vol. 17 Barbara Lewandowska-Tomaszczyk (ed.): Corpus Linguistics, Computer Tools, and Applications – State of the Art. PALC 2007. 2008.

Vol. 18 Barbara Lewandowska-Tomaszczyk / Katarzyna Dziwirek (eds.): Studies in Cognitive Corpus Linguistics. 2009.

Vol. 19 Barbara Lewandowska-Tomaszczyk / Marcel Thelen (eds.): Meaning in Translation. 2010.

Vol. 20 Łukasz Bogucki / Krzysztof Kredens (eds.): Perspectives on Audiovisual Translation. 2010.

Vol. 21 Piotr Stalmaszczyk (ed.): Turning Points in the Philosophy of Language and Linguistics. 2011.

Vol. 22 Stanisław Goźdź-Roszkowski: Patterns of Linguistic Variation in American Legal English. A Corpus-Based Study. 2011.

Vol. 23 Kamila Ciepiela (ed.): Identity through a Language Lens. 2011.

Vol. 24 Stanisław Goźdź-Roszkowski (ed.): Explorations across Languages and Corpora. PALC 2009. 2011.

Vol. 25 Krzysztof Kosecki / Janusz Badio (eds.): Cognitive Processes in Language. 2012.

Vol. 26 Alina Kwiatkowska (ed.): Texts and Minds. Papers in Cognitive Poetics and Rhetoric. 2012.

Vol. 27 Paul A. Wilson (ed.): Dynamicity in Emotion Concepts. 2012.

Vol. 28 Piotr Pęzik (ed.): Corpus Data across Languages and Disciplines. 2012.

Vol. 29 Mikołaj Deckert: Meaning in Subtitling. Toward a Contrastive Cognitive Semantic Model. 2013.

Vol. 30 Łukasz Bogucki: Areas and Methods of Audiovisual Translation Research. 2013.

Vol. 31 Janusz Badio / Kamila Ciepiela (eds.): Language, Action, Interaction. 2013.

Vol. 32 Barbara Lewandowska-Tomaszczyk / Krzysztof Kosecki (eds.): Time and Temporality in Language and Human Experience. 2014.

Vol. 33 Iwona Witczak-Plisiecka (ed.): Cognitive and Pragmatic Aspects of Speech Actions. 2014.

Vol. 34 Małgorzata Anna Kamińska: A History of the *Concise Oxford Dictionary*. 2014.

Vol. 35 Łukasz Bogucki / Mikołaj Deckert (eds.): Accessing Audiovisual Translation. 2015.

Vol. 36 Joanna Jabłońska-Hood: A Conceptual Blending Theory of Humour. Selected British Comedy Productions in Focus. 2015.

Vol. 37 Krzysztof Kosecki / Janusz Badio (eds.): Empirical Methods in Language Studies. 2015.

www.peterlang.com

www.ingramcontent.com/pod-product-compliance
Ingram Content Group UK Ltd.
Pitfield, Milton Keynes, MK11 3LW, UK
UKHW041823031026
15587UKWH00002B/529